P9-EKS-246

Ernst Schering Research Foundation Workshop 35
Stem Cell Transplantation and Tissue Engineering

Springer
Berlin
Heidelberg
New York
Barcelona
Hong Kong
London
Milan
Paris
Tokyo

Ernst Schering Research Foundation Workshop 35

Stem Cell Transplantation and Tissue Engineering

A. Haverich, H. Graf
Editors

With 26 Figures

Springer

Series Editors: G. Stock and M. Lessl

ISSN 0947-6075
ISBN 3-540-41495-9 Springer-Verlag Berlin Heidelberg New York

Die Deutsche Bibliothek - CIP-Einheitsaufnahme

Stem cell transplantation and tissue engineering / A. Haverich and H. Graf ed.. - Berlin ; Heidelberg ; New York ; Barcelona ; Hong Kong ; London ; Milan ; Paris ; Singapore ; Tokyo : Springer, 2002 (Ernst Schering Research Foundation Workshop ; 35) ISBN 3-540-41495-9

This work is subject to copyright. All rights are reserved, whether the whole or part of the material is concerned, specifically the rights of translation, reprinting, reuse of illustrations, recitation, broadcasting, reproduction on microfilms or in any other way, and storage in data banks. Duplication of this publication or parts thereof is permitted only under the provisions of the German Copyright Law of September 9, 1965, in its current version, and permission for use must always be obtained from Springer-Verlag. Violations are liable for prosecution under the German Copyright Law.

Springer-Verlag Berlin Heidelberg New York
a member of BertelsmannSpringer Science+Business Media GmbH

http://www.springer.de

© Springer-Verlag Berlin Heidelberg 2002
Printed in Germany

The use of general descriptive names, registered names, trademarks, etc. in this publication does not imply, even in the absence of a specific statement, that such names are exempt from the relevant protective laws and regulations and therefore free for general use. Product liability: The publishers cannot guarantee the accuracy of any information about dosage and application contained in this book. In every individual case the user must check such information by consulting the relevant literature.

Typesetting: Data conversion by Springer-Verlag
Printing: Druckhaus Beltz, Hemsbach.
Binding: J. Schäffer GmbH & Co. KG, Grünstadt
SPIN: 10793998   21/3130/AG 5 4 3 2 1 0 Printed on acid-free paper

Preface

As dogmas in stem cell research are losing their impact and recent findings regarding the use and cultivation of stem cells and tissue transplantation have opened up new therapeutic avenues, this Ernst Schering Research Foundation Workshop was initiated to highlight current and future approaches in this field.

The only stem cells that have been being used clinically for some time now are hematopoietic stem cells, as a source for bone marrow transplantation. Interestingly, recent findings indicate that hematopoietic stem cells have the ability not only to differentiate into blood cells but also, under certain conditions, into endothelial, neural or muscle cells, providing exciting new therapeutic possibilities. They may represent a future source for tissue engineering, replacing defective cells or tissues and allowing diseased organs to regain their functions .

In addition to hematopoietic stem cells, attention is increasingly being directed to neural stem cells. The focus of this research is adult neurogenesis. As adult neuronal stem cells can apparently be found anywhere in the brain, neurogenesis depends not only on them but also on the microenvironment permitting neurogenesis. The studies presented by Gerd Kempermann clearly indicate that neurogenesis is regulated in an activity-dependent manner and lead to the conclusion that neurogenesis in the adult brain is a physiological property and not as much of an exception as previously thought. These findings might be the basis for new therapeutic strategies for the treatment of degenerative brain diseases such as Parkinson.

One step beyond stem cell transplantation is the "art" of tissue engineering. This can actually be compared to the work of an artist or desi-

The participants of the workshop

gner, as the construction of three-dimensional structures is a complex and difficult process. The goal of tissue engineering is none less than the synthesis of viable, functional tissues for transplantation. The earliest attempts at engineering a tissue mass were carried out by Bell and Yannas at the Massachusetts Institute of Technology in the late 1970s, early 1980s. Their approach relied on the use of collagen-based gels and foams to provide the necessary structural definition for their proliferation and the differentiation of neonatal human foreskin fibroblasts to an epidermis-like tissue. These techniques have come of age and innovative polymers are now being used for the three-dimensional construction of tissues as highlighted by Shastri and Martin in this volume. One of the most intriguing examples is the grafting of a polymer construction seeded with chondrocytes in a nude mouse to generate a tissue in the shape of a human ear.

Due to the fact that cardiovascular diseases and particularly ischemic disorders of the heart are the leading causes of death in the Western world and the average waiting time for heart transplants is between 9 and 12 months in Germany, alternative approaches are necessary. Therefore, the special focus of this workshop was on techniques with which diseased heart cells or blood vessels can be reconstituted. Selbert and Franz report on the successful generation of ventricular-like cardiomyocytes from mouse embryonic stem cells in vitro. The isolated cardiomyocytes show sarcomeric structures, are beating, and are ready for transplantation into the diseased heart. These results can be seen as a first step towards the cultivation of human cardiomyocytes in vitro. In this respect however, ethical considerations regarding the use of human embryonic stem cells have to be taken into account.

In the laboratory of Francois Auger a very elegant technique for the formation of tissue-engineered blood vessels has been developed and is described in this book. It is based exclusively on the use of human cells in the absence of any synthetic or exogenous material such as animal collagens. This vessel was shown to have a supra-physiological blood pressure resistance and a histological organization comparable to that of a native artery. Thus, the tissue-engineered blood vessels offer exciting perspectives in both clinical and pharmacological applications.

Taken together, the progress that has been made in the understanding of stem cell biology, on the one hand, and in the "art" of tissue engineering, on the other, is striking. Both reconstitution techniques are paving the way for the development of new therapeutic strategies, giving hope of being able to cure and not only treat patients.

M. Lessl

Contents

List of Editors and Contributors

Editors

A. Haverich
Klinik für Thorax-, Herz- und Gefäßchirurgie, Medizinische Hochschule
Hannover, Germany

H. Graf
Strategic Business Unit Therapeutics, Schering AG, Müllerstr. 178,
13342 Berlin, Germany

Contributors

Atala, A.A.
Laboratory for Tissue Engineering and Cellular Therapeutics,
Children's Hospital and Harvard Medical School, 300 Longwood Avenue, Boston, MA 02115, USA

Auger, F.A.
Laboratoire d'Organogénèse Expérimentale/LOEX, Hôpital du Saint-Sacrement and Department of Surgery, Faculty of Medicine, Laval University,
Québec City, Québec G1S 4L8, Canada

Bader, A.
Gesellschaft für Biotechnologische Forschung Braunschweig (GBF), Organ
und Gewebekulturen, Mascheroder Weg 1, 38124 Braunschweig, Germany

De Bartolo, L.
Research Institute on Membranes and Modelling of Chemical Reactors, IR-
MERC-CNR, c/o University of Calabria, via P. Bucci, cubo 17/C,
87030 Rende (CS), Italy

Franz, W.
Medizinische Universitätsklinik, Medizinische Klinik II,
Ratzeburger Allee 100, 23538 Lübeck, Germany

Germain, L.
Laboratoire d'Organogénèse Expérimentale/LOEX, Hôpital du Saint-Sacre-
ment and Department of Surgery, Faculty of Medicine, Laval University,
Québec City, Québec G1S 4L8, Canada

Grenier, G.
Laboratoire d'Organogénèse Expérimentale/LOEX, Hôpital du Saint-Sacre-
ment and Department of Surgery, Faculty of Medicine, Laval University,
Québec City, Québec G1S 4L8, Canada

Kempermann, G.
AG Neuronale Stammzellen, Max Delbrück Centrum für Molekulare Medi-
zin, Robert-Rössle-Strasse 10, 13125 Berlin, Germany

Martin, I.
Department of Surgery, Research Division, Kantonsspital Basel, Switzerland

McKay, R.
NINDS, Mol Biol Lab, NIH, 36 Convent Drive, Building 36, Room 5A29,
Bethesda, MD 20892, USA

Menasché, P.
Department of Cardiovascular Surgery, Hôpital Bichat Claude Bernard,
46, rue Henri Huchard, 75018 Paris, France

Ostermann, H.
Medizinische Klinik und Poliklinik III, LMU Klinikum Großhadern,
Marchioninistrasse 15, 81377 München, Germany

Rémy-Zolghadri, M.
Laboratoire d'Organogénèse Expérimentale/LOEX, Hôpital du Saint-Sacrement and Department of Surgery, Faculty of Medicine, Laval University, Québec City, Québec G1S 4L8, Canada

Selbert, S.
Universitätsklinikum Lübeck, Medizinische Klinik II, Ratzeburger Allee 160, 23538 Lübeck, Germany

Shastri, P.
Department of Materials Science and Engineering and School of Medicine, University of Pennsylvania, Abramson Pediatric Research Center, Suite 707C, 34th St. & Civic Center Blvd., Philadelphia, PA 19104

Yoo, J.S.
Laboratory for Tissue Engineering and Cellular Therapeutics, Children's Hospital and Harvard Medical School, 300 Longwood Avenue, Boston, MA 02115, USA

1 Bone Marrow Reconstitution

H. Ostermann

1 Introduction

An observation 50 years ago – rescue of mice from lethal irradiation by shielding the spleen – rapidly led to the clinical use of bone marrow transplantation (BMT). It was observed that reconstitution of the bone marrow took place from cells that could be found in the non-irradiated spleen. Shortly thereafter it was shown that bone marrow taken from one mouse could reconstitute the bone marrow in a lethally irradiated mouse. These observations and further canine experiments led to the clinical concept of BMT (Thomas 1999a,b). The approach of using BMT for the treatment of severely ill patients was achieved before the cells responsible for the reconstitution of the bone marrow had been identified and characterised. Nevertheless, a rapid transition from animal experiments to clinical practice was necessary because the first use of BMT was in patients who had no other chance of cure. These were patients with relapsed or refractory leukaemia who would otherwise have died in a short time. However, it could be proved that by BMT some of these patients had a chance of being cured.

2 The Haematopoietic Stem Cell

The haematopoietic stem cell (HSC) is traditionally regarded as a cell that resides in the bone marrow and is able to reconstitute all cell lines involved in haematopoiesis (erythrocytes, thrombocytes, leukocytes and lymphocytes). In addition to its occurrence in the bone marrow as evidenced by the successful transplantation of bone marrow, the HSC has some unique properties that differentiate it from the other organ-related stem cells that are discussed at this meeting. The HSC is able to leave the bone marrow and repopulate the bone marrow at a site that is distinct from its original place of residence. As such, HSCs can be found in the circulating blood. It is well known that in the phase of regeneration following chemotherapy or after the application of haematopoietic growth factors, a dramatic increase of these stem cells can be observed in the circulation, making it possible to collect HSCs for therapeutic approaches (Tarella et al. 1995; Olavarria and Kanfer 2000). The HSC has been well characterised on morphological, immunological, functional and genetic grounds. Furthermore, a substantial body of clinical evidence and experience has been gained by using these stem cells in clinical transplantation for about 30 years.

2.2 Characterisation

The HSC has been characterised by several means within the last several years. Morphologically the stem cell resembles lymphocytes and can thus not be distinguished easily by microscopy. However, progress in the functional characterisation of the stem cell as well as the definition of a distinct phenotype by analysing the expression of surface antigens and the genetic repertoire has helped us substantially to understand the characteristics of the HSC.

2.3 Phenotype

It was recognised early that HSCs carry a distinct set of surface antigens. The most commonly accepted characterisation of HSCs includes the co-expression of CD34 and CD90 (Thy-1) together with the absence of those antigens that mark lineage specificity like CD38 (Bhatia et al. 1997). The exact function of the "stem cell antigens" CD34 and CD90 is unknown, but it is suggested that CD34 is an adhesion molecule, while CD90 has been implicated in signal transduction. It is clear, however, that the CD34+, CD90+, lineage-negative population is still a heterogeneous population of cells. The "real" human stem cell has not been defined yet; however, existing data support the notion that the stem cell could reside within a CD34-negative population. Evidence from mice experiments indicates that using single, CD34-negative stem cells, repopulation of lethally irradiated mice is possible (Osawa et al. 1996). Furthermore, CD34-negative, lineage-negative human cells were able to engraft in non-obese/severe combined immunodeficiency (NOD/SCID) mice, showing for the first time haematopoietic activity from the CD34-negative population (Bhatia et al. 1998). The frequency of repopulating cells within the CD34-negative, lineage-negative compartment seems to be lower than the number of repopulating cells in the CD34-positive population. Further characterisation of the repopulating cells from the CD34-negative fraction have revealed a subset that is positive for AC133, lacking CD7 and CD34, and is lineage negative. This subset has a frequency of only 0.2% within the CD34-CD38-lineage fraction, however, this population has progenitor capacity, is the only fraction within the CD34-CD38-Lin fraction that is able to generate CD34+ cells and it is able to repopulate NOD/SCID mice (Gallacher et al. 2000). Thus, further progress in identifying the ultimate HSC is made. These data will finally allow the selection of very pure HSCs from bone marrow or peripheral blood for autologous or allogeneic transplantation, as well as targets for gene transfer.

2.4 Development from Stem Cell to Mature Blood Cell

The HSC is characterised by the ability to generate all haematopoietic cell lines (Fig. 1). Thus red cells, platelets, lymphocytes, granulocytes and monocytes can all be generated from a single HSC. It has clearly been shown for lymphocytes and for myeloid cells that a common progenitor is able to generate all these differentiated cells (Osawa et al. 1996; Akashi et al. 1999, 2000). These abilities led to a model of haematopoietic maturation that characterises the developmental stages

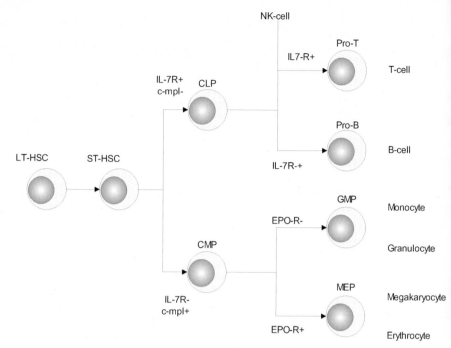

Fig. 1. Model for major haematopoietic pathways from HSCs. Long-term HSCs (LT-HSCs) give rise to short-term HSCs (ST-HSCs). The ST-HSCs give rise to common lymphoid progenitors (CLPs) and common myeloid progenitors (CMPs). These differentiate into the T-/B-cell lymphocyte lineage and the monocyte/granulocyte–megakaryocyte/erythrocyte lineage respectively. Each important step in this differentiation process can be defined by a specific expression of genes as depicted. From Akashi et al. (2000)

by the acquisition or loss of certain surface markers or by the expression of distinct genes during the development (Phillips et al. 2000). Thus a common HSC can be defined which gives rise to either lymphoid or myeloid progenitors. The myeloid progenitor pathway differentiates again into the monocyte/granulocyte and the megakaryocyte/erythrocyte lineage (Akashi et al. 2000).

2.5 Stem Cell Assays

As the human stem cell has not yet been definitely identified, assays for stem cells must rely on methods which identify the population associated with stem cells. As depicted above, the phenotype as detected by the expression of surface antigens is important in identifying early progenitor cells. In clinical practice, the expression of the CD34 antigen has been used to characterise the progenitor content of leukapheresis products. Although it is probable that the human HSC could be CD34 negative, the reconstitution of the bone marrow following HSC transplantation after myeloablative therapy correlates with the amount of CD34-positive cells given. Thus, CD34 positivity is a useful clinical marker for HSC. However, the phenotype that is probably most related to HSC can be defined as CD34-CD38-Lin-CD7-AC133$^+$ (Gallacher et al. 2000).

Other assays have been developed which rely on the ability of HSC to form colonies which comprise the different haematopoietic lineages. These assays are being described as assaying activity rather closer to the HSC, like the cobblestone area forming cell (CAFC), or the long-term culture initiating cell (LTCIC), or reflecting the presence of early progenitor cells like the colony-forming units (CFU) assays. CFU assays show a correlation to CD34 counts and to haematopoietic recovery following myeloablative treatment, thus they have been adopted for clinical use. CAFC and related assays are time-consuming (taking up to 4 weeks to be read) and have thus not had clinical impact.

3 Clinical Use of Stem Cell Transplantation

3.1 Allogeneic Vs Autologous Transplantation

Historically the transplantation of stem cells was first performed in the allogeneic setting in patients with refractory leukaemia (Thomas 1999b). Allogeneic transplantation was performed following myeloablative treatment of the recipient with the idea of eradicating the leukaemia cells in the bone marrow. Simultaneously, the normal haematopoiesis was destroyed. In order to live, the patient had to receive a histocompatible marrow graft (Heslop 1999) from either a matched sibling or an HLA-identical unrelated donor (Stroncek et al. 2000). At the time it was thought that the main reason why patients could be cured with such an approach was the complete eradication of the malignant cells by the myeloablative treatment. Recently, however, it could be shown that myeloablation is not necessary to cure patients (Khouri et al. 1998). It is sufficient to suppress the immune system of the recipient to such a degree that the graft will not be rejected. The newly implanted graft will then not only provide haematopoiesis but also reconstitute the immune system. The new immune system is thought to be a very important aspect of the treatment regimen since it regards the leukaemic cells as immunologically different and destroys them. Thus, the concept of myeloablation is becoming replaced by a concept of a combined approach of lympho- and myeloablation.

3.2 The Source of Stem Cells

Historically the source for stem cells was bone marrow. Animal experiments had shown that transplantation was possible using bone marrow, thus it was used in patients as well (Thomas 1999b). However, circulating cells able to repopulate a recipient have been shown to exist. They circulate in steady state; however, the number of these cells is too low to be collected for therapeutic reasons. Following chemotherapy, it has been observed that the recovery of peripheral blood cells after leukocytopenia is accompanied by an increase in circulating progenitor cells. These can be collected by leukapheresis. If chemotherapy is combined with the application of growth factors, the yield of progenitor cells can be further increased (Stewart et al. 1999). It is usually sufficient to perform 1 or 2 leukapheresis procedures to obtain an amount of progenitor cells suitable for transplantation. The surrogate marker for the amount of stem cells in the leukapheresis product is the number of $CD34^+$ cells. A dose of 2×10^6/kg is sufficient for repopulating bone marrow after myeloablation; however, higher doses in the autologous and in certain aspects in the allogeneic setting may be beneficial (Table 1) (Siena et al. 2000; Powles et al. 2000).

Table 1. Mobilisation of progenitor cells by growth factors and chemotherapy. The use of granulocyte colony-stimulating factor (G-CSF) alone results in a modest yield of progenitor cells. This can be increased if a disease-orientated chemotherapy is followed by G-CSF, or even more if intensive chemotherapy is used. From Stewart et al. (1999)

Mobilisation scheme	G-CSF alone	Chemotherapy +G-CSF	Intensive chemotherapy
CD34$^+$×10^6/l	21.5	65	278
CD34$^+$/kg	3.3	6.9	18.3
Apheresis volume (l)	25.1	15.4	11.3

3.3 Selection and Purging of Stem Cells

In the autologous setting, the patient's marrow or peripheral blood are the source for stem cells to be used following myeloablative treatment. Involvement of the bone marrow occurs inherently in some diseases like leukaemias, in others, like lymphoma or solid tumours, it often is found. Thus, collecting bone marrow or peripheral blood stem cells from these patients might yield an apheresis product that could be contaminated with tumour cells. These tumour cells might then be the cause of relapse from the disease (Brenner et al. 1994). Thus methods have been developed to purge tumour cells from the apheresis product or bone marrow. They can basically be divided into those methods that enrich CD34$^+$ cells (CD34 selection) and those methods that are used to eliminate tumour cells from the stem cell product (purging) (Vogel et al. 2000).

3.3.1 CD34 Selection
The selection of CD34 cells in apheresis products can be performed using immunomagnetic beads which are added to anti-CD34-sensitised progenitor cells. Either positive selection for progenitor cells or negative selection to eliminate tumour cells from the graft can be performed. It has been shown that it is possible to safely treat patients using these cells, as prompt recovery of haematopoiesis does occur following transplantation of purged progenitor cells (Vescio et al. 1999). The results as to the outcome of transplantation with selected cells so far have been ambiguous. Most studies employing these techniques have not shown that using CD34-selected progenitor cells provide an advantage over

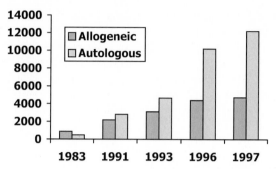

Fig. 2. Development of allogeneic vs autologous transplantations in Europe

non-selected cells. The reasons for these so far disappointing results could be that either the selection process is not efficient enough, leaving residual tumour cells in the graft, or that the high-dose therapy is not sufficient to eradicate all residual tumour cells in the host.

3.4 Indications for Stem Cell Transplantation

The clinical observation that patients with advanced leukaemia who only had a short period of time to live using conventional treatment could be cured using allogeneic bone marrow led to the rapid introduction of allogeneic BMT into clinical practice (Fig. 2; Thomas 1999a,b).

The concept of intensifying the dosage of chemotherapy to have a more efficient way to treat patients with malignancies led to problems arising from the myelosuppressive side-effects of chemotherapy. The use of autologous bone marrow was shown to lead to a rapid recovery following intensive chemotherapy.

This paved the way for clinical studies in the autologous setting. Meanwhile, some indications have emerged for the use of autologous transplants following high-dose therapy in patients with malignancies.

3.4.1 Autologous

Dose intensification has been shown to be of value in the treatment of some malignancies. The concept behind this approach is the idea that if a certain amount of tumour cells can be killed by a certain amount of chemotherapy then it should be possible by increasing the dose of cytotoxic drugs to increase the efficacy of the treatment. However, most chemotherapy regimens have been tailored according to their toxicities to safely treat patients. A further increase in dosage would thus lead to an unacceptable increase in toxicity. One major source of toxicity with chemotherapeutic drugs is myelosuppression. As a prolonged period of myelosuppression is associated with a high risk of serious infections and bleeding complications, or even myeloablation, a way to overcome these complications has been the idea of autologous transplantation. Within the last 10 years the number of autologous transplantations has increased substantially (Gratwohl et al. 1997).

3.4.2 Lymphoma

Patients with lymphomas have been thoroughly evaluated for the efficacy of high-dose therapy followed by progenitor cell transplantation. The reason for this is the high degree of chemotherapy sensitiveness of this disease. However, only about 30%–50% of patients with high-grade lymphoma and only a few patients with low-grade lymphoma can be cured using an approach of conventional-dose chemotherapy. Thus, high-dose therapy was evaluated in these patients in an effort to increase the chances of cure. Among the plethora of studies that have been published in patients with lymphomas, a few should be mentioned.

Patients with aggressive lymphoma in chemosensitive relapse were randomised to receive either further conventional chemotherapy or high-dose therapy with autologous BMT (Philip et al. 1995). Since a significant advantage in 5-year overall survival for transplant patients was found, in patients with sensitive relapse from high-grade lymphoma, high-dose therapy followed by progenitor cell transplantation is now be regarded as standard treatment in suitable patients. In patients with slowly responding or primary refractory high-grade non-Hodgkin's lymphoma (NHL), the evidence for the benefits of high-dose therapy is less clear (Verdonck et al. 1995; Martelli et al. 1996). In first remission following standard therapy, an advantage for high-dose therapy has been reported (Haioun et al. 2000; Gianni et al. 1997).

In follicular lymphoma as the predominant low-grade NHL, randomised trials have not been published; however, in relapsed follicular-lymphoma NHL, overall survival of 66% at 8 years has been reported (Freedman et al. 1999). Purging of the graft may be a useful approach in follicular lymphoma.

3.4.3 Multiple Myeloma

As in low-grade NHL in multiple myeloma, cure with conventional chemotherapy is not possible. In a randomised trial by the Intergroup Français du Myélome published in 1996, untreated patients with multiple myeloma were randomised to conventional chemotherapy alone or high-dose chemotherapy followed by autologous BMT (Attal et al. 1996). An 81% response rate in transplanted patients compared to a 57% response rate in patients treated with chemotherapy only was reported. The estimated rate of survival was 32% vs 12% ($p=0.03$). An alternative approach termed "total therapy" by the Little Rock group used a further intensification of the chemotherapy preceding the high-dose therapy. In patients achieving a complete remission (CR) the median duration of CR was an impressive 50 months (Barlogie et al. 1999). However, a plateau has not been reached in any of the studies on high-dose therapy in multiple myeloma, suggesting that cure may not be achievable in patients treated with high-dose therapy. Although the occurrence of plasma cells in bone marrow or leukapheresis products is a common observation, a positive effect of selection for CD34 positive cells has not been convincingly shown (Vescio et al. 1999).

3.4.4 Acute Myelogenous Leukaemia

Although the majority of patients with acute myelogenous leukaemia (AML) achieve a complete remission following induction chemotherapy, many will eventually relapse and die. A randomised study has shown that post-remission therapy with high-dose therapy and progenitor cell transplantation is superior to one more courses of conventional chemotherapy (Burnett et al. 1998). Although the post-remission treatment of choice in patients with AML achieving a first CR is transplantation from an HLA-identical family member, autologous transplantation following high-dose therapy is a viable alternative.

3.4.5 Solid Tumours

Although the approach of high-dose therapy and progenitor cell transplantation has been investigated in a lot of solid tumours, patients with breast cancer have been most often treated. The use of high-dose therapy in breast cancer has, however, rapidly declined since it became obvious last year that the investigator who had published two positive randomised trials of high-dose therapy in adjuvant and in metastatic breast cancer, both with a significant positive result, was charged with fraud. In the meantime, there have been some negative studies (Stadtmauer et al. 2000); however, the final results of some large prospective randomised trials have not yet been published and the issue should thus be kept debatable at least until the publication of these trials.

3.4.6 Autoimmune Disease

Autoimmune diseases are a heterogeneous group of diseases. As the aetiology is unknown, a causal therapy is not possible. As the prognosis of some forms of autoimmune disease is, however, dismal, a search for innovative approaches is warranted. Immune suppression is the mainstay of treatment in autoimmune diseases. Since high-dose therapy followed by progenitor cell transplantation is an intensely immunosuppressive treatment, it was evaluated in several patient groups with autoimmune diseases (Jantunen et al. 2000; Marmont 2000; Traynor et al. 2000). However, as organ involvement, for example of heart and kidneys, is often observed in these patients, the complication rate following high-dose therapy may be higher than in patients with malignancies. Randomised trials are now underway to provide evidence for this new therapeutic approach.

3.4.7 Allogeneic

Allogeneic transplantation from a HLA-identical sibling or a matched unrelated donor has been performed for more than 40 years. Randomised trials comparing allogeneic transplantation are generally not available; however, this is mostly due to the fact that in most patients groups, i.e. refractory acute leukaemia, on ethical grounds a randomised trial cannot be performed. The positive effect of allogeneic transplantation resulting in a high rate of relapse-free remissions is clouded by the fact that acute and chronic complications do occur in a substantial number of patients resulting in a treatment associated mortality rate of about 20%–40%, depending on the kind of diseases, the number and kind of previous treatments and various patient factors like age and Karnofsky index.

While in the autologous setting the majority of transplant patients have solid tumours or lymphomas, the mainstays of allogeneic transplantation are patients with leukaemia and those with a life-threatening

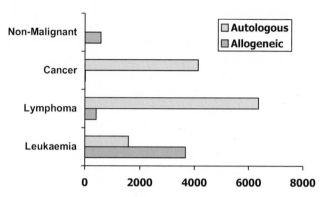

Fig. 3. Indications for allogeneic and autologous transplantations in Europe (Gratwohl et al. 1999)

non-malignant disease like sickle cell anaemia or hereditary disorders of metabolism (Fig. 3).

The source of stem cells in allogeneic transplantation has traditionally been bone marrow. However, in recent years a shift towards the use of peripheral blood stem cells has occurred. It could be shown that transplantation is safe using peripheral blood progenitor cells (PBPCs); haematological recovery occurs substantially faster than with bone marrow and complications like graft-versus-host disease or even relapse do not seem to be substantially different (Stroncek et al. 2000; Powles et al. 2000). In order to reduce the probability of graft rejection and of graft-versus-host disease, a fully HLA-matched donor should be available. In the family setting, a sibling has a 25% chance of sharing both HLA haplotypes with the donor. If a suitable family donor is not available, the existing worldwide donor registries can be searched for a matched unrelated donor. The chances of finding a suitable donor are, at least within the white Caucasian population, good. Despite these efforts, not every patient will be expected to have a suitable donor. Recently, new strategies have been developed to overcome this obstacle. One way is to use fully mismatched HLA haplotype donors. Using this approach, nearly every patient will have a donor available in his or her family. To ensure safe engraftment with low complication rates, a strong immuno-suppressive conditioning regimen has to be used. In combination with the use of peripheral blood progenitor cells which were T-cell depleted, this approach achieved stable engraftment with few complications (Aversa et al. 1998).

Toxicity of the conditioning regimen can be reduced by using a more immunosuppressive approach compared to the myeloablative approach. While engraftment is possible in most patients using these measures, the effect of the graft-versus-tumour effect becomes more obvious in this setting (Khouri et al. 1998). It has been known for some time that relapse following BMT can be reverted to remission again purely by the application of donor-derived lymphocytes (Kolb et al. 1990). This effect is immune-cell-mediated against the leukaemia in the recipient. Translating these results to the observed possibility to transplant with a non-myeloablative regimen, it becomes obvious that a major, if not the principle, part of treatment is the immune-mediated anti-tumour approach. That myeloablation is not the main factor in transplantation is also evident from the fact hat in autologous transplantation, cure rates are always substantially lower than in allogeneic transplantation, even if a myeloablative regimen is used in autologous transplantation. Reducing the intensity of the conditioning regimen does also mean reducing the regimen-related toxicities, thus allowing older patients or those with co-morbidities which are prohibitive for myeloablative treatment to be transplanted.

4 Outlook: Stem Cells from Bone Marrow

The recent discovery that stem cells found in bone marrow have the ability to not only differentiate into blood cells but also, under certain conditions, differentiate into endothelial cells, neural cells or muscle cells provides exciting possibilities for stem cell research (Devine and Hoffman 2000). These mesenchymal stem cells may provide a future source for tissue engineering and may be able to replace defective cells or tissues allowing us to regain function in diseased organs. Furthermore, autologous mesenchymal stem cells may be an appropriate target for gene therapy, providing it is possible to specifically induce differentiation in these stem cells and provide organ specificity.

References

Akashi K, Traver D, Kondo M, Weissman IL (1999) Lymphoid development from hematopoietic stem cells. Int J Hematol 69:217–226

Akashi K, Traver D, Miyamoto T, Weissman IL (2000) A clonogenic common myeloid progenitor that gives rise to all myeloid lineages. Nature 404:193–197

Attal M, Harousseau JL, Stoppa AM et al (1996) A prospective, randomized trial of autologous bone marrow transplantation and chemotherapy in multiple myeloma. Intergroupe Francais du Myelome. N Engl J Med 335:91–97

Aversa F, Tabilio A, Velardi A et al (1998) Treatment of high risk acute leukemia with T-cell-depleted stem cells from related donors with one fully mismatched HLA haplotype. N Engl J Med 339:1186–1193

Barlogie B, Jagannath S, Desikan KR et al (1999) Total therapy with tandem transplants for newly diagnosed multiple myeloma. Blood 93:55–65

Bhatia M, Bonnet D, Murdoch B, Gan OI, Dick JE (1998) A newly discovered class of human hematopoietic cells with SCID- repopulating activity (see comments). Nat Med 4:1038–1045

Bhatia M, Wang JC, Kapp U, Bonnet D, Dick JE (1997) Purification of primitive human hematopoietic cells capable of repopulating immune-deficient mice. Proc Natl Acad Sci USA 94:5320–5325

Brenner M, Krance R, Heslop HE et al (1994) Assessment of the efficacy of purging by using gene marked autologous marrow transplantation for children with AML in first complete remission. Hum Gene Ther 5:481–499

Burnett AK, Goldstone AH, Stevens RM et al (1998) Randomised comparison of addition of autologous bone-marrow transplantation to intensive chemotherapy for acute myeloid leukaemia in first remission: results of MRC AML 10 trial. UK Medical Research Council Adult and Children's Leukaemia Working Parties. Lancet 351:700–708

Devine SM, Hoffman R (2000) Role of mesenchymal stem cells in hematopoietic stem cell transplantation. Curr Opin Heamatol 7:358–363

Freedman AS, Neuberg D, Mauch P et al (1999) Long-term follow-up of autologous bone marrow transplantation in patients with relapsed follicular lymphoma. Blood 94:3325–3333

Gallacher L, Murdoch B, Wu DM, Karanu FN, Keeney M, Bhatia M (2000) Isolation and characterization of human CD34(-)Lin(-) and CD34(+)Lin(-) hematopoietic stem cells using cell surface markers AC133 and CD7. Blood 95:2813–2820

Gianni AM, Bregni M, Siena S et al (1997) High-dose chemotherapy and autologous bone marrow transplantation compared with MACOP-B in aggressive B-cell lymphoma. N Engl J Med 336:1290–1297

Gratwohl A, Passweg J, Baldomero H, Hermans J (1999) Blood and marrow transplantation activity in Europe 1997. European Group for Blood and Marrow Transplantation (EBMT). Bone Marrow Transplant 24:231–245

Haioun C, Lepage E, Gisselbrecht C et al (2000) Survival benefit of high-dose therapy in poor-risk aggressive non-Hodgkin's lymphoma: final analysis of the prospective LNH87-2 protocol– a groupe d'Etude des lymphomes de l'Adulte study. J Clin Oncol 18:3025–3030

Heslop HE (1999) Haemopoietic stem cell transplantation from unrelated donors. Br J Haematol 105:2–6

Jantunen E, Myllykangas-Luosujarvi R (2000) Stem cell transplantation for treatment of severe autoimmune diseases: current status and future perspectives. Bone Marrow Transplant 25:351–356

Khouri IF, Keating M, Korbling M et al (1998) Transplant-lite: induction of graft-versus-malignancy using fludarabine-based nonablative chemotherapy and allogeneic blood progenitor-cell transplantation as treatment for lymphoid malignancies. J Clin Oncol 16:2817–2824

Kolb HJ, Mittermüller J, Clemm C, Holler E, Ledderose G, Brehm G, Heim M, Wilmanns W (1990) Donor leukocyte transfusions for treatment of recurrent chronic myelogenous leukemia in marrow transplant patients. Blood 76(12):2462–2465

Marmont AM (2000) New horizons in the treatment of autoimmune diseases: immunoablation and stem cell transplantation. Annu Rev Med 51:115–134

Martelli M, Vignetti M, Zinzani PL et al (1996) High-dose chemotherapy followed by autologous bone marrow transplantation versus dexamethasone, cisplatin, and cytarabine in aggressive non-Hodgkin's lymphoma with partial response to front-line chemotherapy: a prospective randomized Italian multicenter study. J Clin Oncol 14:534–542

Olavarria E, Kanfer EJ (2000) Selection and use of chemotherapy with hematopoietic growth factors for mobilization of peripheral blood progenitor cells. Curr Opin Hematol 7:191–196

Osawa M, Hanada K, Hamada H, Nakauchi H (1996) Long-term lymphohematopoietic reconstitution by a single CD34- low/negative hematopoietic stem cell. Science 273:242–245

Philip T, Guglielmi C, Hagenbeek A et al (1995) Autologous bone marrow transplantation as compared with salvage chemotherapy in relapses of chemotherapy-sensitive non-Hodgkin's lymphoma. N Engl J Med 333:1540–1545

Phillips RL, Ernst RE, Brunk B et al (2000) The genetic program of hematopoietic stem cells. Science 288:1635–1640

Powles R, Mehta J, Kulkarni S et al (2000) Allogeneic blood and bone-marrow stem-cell transplantation in haematological malignant diseases: a randomised trial. Lancet 355:1231–1237

Siena S, Schiavo R, Pedrazzoli P, Carlo-Stella C (2000) Therapeutic relevance of CD34 cell dose in blood cell transplantation for cancer therapy. J Clin Oncol 18:1360–1377

Stadtmauer EA, O'Neill A, Goldstein LJ et al (2000) Conventional-dose chemotherapy compared with high-dose chemotherapy plus autologous hematopoietic stem-cell transplantation for metastatic breast cancer. Philadelphia Bone Marrow Transplant Group. N Engl J Med 342:1069–1076

Stewart DA, Guo D, Morris D et al (1999) Superior autologous blood stem cell mobilization from dose-intensive cyclophosphamide, etoposide, cisplatin plus G-CSF than from less intensive chemotherapy regimens. Bone Marrow Transplant 23:111–117

Stroncek DF, Confer DL, Leitman SF (2000) Peripheral blood progenitor cells for HPC transplants involving unrelated donors. Transfusion 40:731–741

Tarella C, Benedetti G, Caracciolo D et al (1995) Both early and committed haemopoietic progenitors are more frequent in peripheral blood than in bone marrow during mobilization induced by high-dose chemotherapy + G-CSF. Br J Haematol 91:535–543

Thomas ED (1999a) Bone marrow transplantation: a review. Semin Hematol 36:95–103

Thomas ED (1999b) A history of haematopoietic cell transplantation. Br J Haematol 105:330–339

Traynor AE, Schroeder J, Rosa RM et al (2000) Treatment of severe systemic lupus erythematosus with high-dose chemotherapy and haemopoietic stem-cell transplantation: a phase I study. Lancet 356:701–707

Verdonck LF, van Putten WL, Hagenbeek A et al (1995) Comparison of CHOP chemotherapy with autologous bone marrow transplantation for slowly responding patients with aggressive non-Hodgkin's lymphoma. N Engl J Med 332:1045–1051

Vescio R, Schiller G, Stewart AK et al (1999) Multicenter phase III trial to evaluate CD34(+) selected versus unselected autologous peripheral blood progenitor cell transplantation in multiple myeloma. Blood 93:1858–1868

Vogel W, Scheding S, Kanz L, Brugger W (2000) Clinical applications of CD34(+) peripheral blood progenitor cells (PBPC). Stem Cells 18:87–92

2 Neuronal Stem Cells and Adult Neurogenesis

G. Kempermann

1 Stem Cell Biology and Modern Medicine

In many fields of modern medicine and medically oriented biology the growing interest in stem cells has fundamentally changed the perception of what is therapeutically possible. In principle, stem cell biology has introduced cellular replacement strategies even to fields where classical organ transplantation, such as heart and kidney, was impossible – most notably neurobiology (Gage 2000).

Stem cells are undifferentiated cells with the potential for various lineages of cellular differentiation. The ultimate stem cell, which is totipotent in that it can give rise to all body tissues, is the fertilized egg (Donovan 1998). Stem cells do not differentiate themselves, but have the capacity of self-renewal through cell division. The current concept is that these divisions can be either symmetric, generating two new stem cells with essentially the same potency as the mother cell, or asymmetric, resulting in one identical copy and one different cell with a more limited spectrum of potential lineages of differentiation. Not unlike

Greek mythology, there is a price to pay for leaving the eternal spheres: the offspring of the "eternal" stem cells becomes increasingly restricted in their potency and lifespan. At least this is part of the prevailing theory. As a consequence, "stem cells" exist on numerous levels of organ development, from totipotent egg to pluripotent or multipotent, tissue-specific stem cells. Tissue-specific stem cells persist throughout adulthood and provide a regenerative potential for certain organs. Epidermal cells, for example, continuously generate new keratocytes, resulting in a high cellular turnover of the outer body surface and a large regenerative capacity after bruises, cuts and even burns.

The term "stem cell" is applied to a variety of cell types that highly differ in character, function and potential. This heterogeneity burdens the discussion about practical and ethical implications of stem cell-based therapies (or experiments).

Embryonic stem (ES) cells are derived from blastocysts in the first days of embryonic development. They are "almost" totipotent, but not quite, because unlike a single fertilized egg cell, a single embryonic stem cell cannot develop into a complete organism on its own. However, within the right cellular context ES cells can differentiate into probably all types of body cells (including tissue-specific stem cells), making them some sort of "universal donor" for stem-cell based cellular replacement strategies (Brüstle et al. 1998).

Widely used in murine knock-out technology, ES cells sparked considerable ethical and legal discussions after human ES cells were first used in experiments because human ES cells have to be, as their name implies, derived from human embryos (Thomson and Odorico 2000). Therefore, while their potential is vast, human ES cells are likely to underlie complex restrictions in their potential experimental and clinical application.

Embryonic germ line (EG) cells seem to be very similar to ES cells (Shamblott et al. 1998; Thomson et al. 2000). But they are derived bioptically from the developing reproductive organs of an embryo, potentially without doing harm to the embryo. While some hail EG cells as the chance to escape from the dilemma surrounding ES cells, others maintain that manipulation of an EG cell still constitutes a ethically problematical manipulation of an embryo, no matter if the rest of the developing embryo is spared or not.

Although (and as so often in science and the perception of science in the public), the ethical and legal issues around ES and EG cells are far more complex and ambiguous than they might seem at first sight; it is fairly obvious that research in this field will be widely reigned and influenced by public debate, fears and consideration beyond the research and its results.

2 Complicating Factors

In a way, the problems regarding stem cells go right to the heart of the debate about Dolly, the first cloned sheep (Wilmut et al. 1997). How much of the individual is in a single cell? The discussion about cloning has already greatly influenced the perception of "stem cell biology", and not only in the public. However, much of stem cell biology does not even get near cloning techniques and problems, although the overlap exists. Recent data from Jonas Frisén's laboratory at the Karolinska Institute in Stockholm have pushed this ambiguity one step further. This group showed that stem cells derived from the adult brain could be "reprogrammed" by contact with an embryonic microenvironment to act like embryonic stem cells (Clarke et al. 2000). This implies that one might not generally have to touch any embryo to generate embryonic stem cells. While most of the public has just started to realize the ethical and legal issues around embryonic stem cells (in the "old" sense of the word), the bases for hopes, opportunities and ethical questions have already substantially shifted and will by all means change the path of future discussion. The loss of the sharp distinction between embryonic and adult stem cells, and thereby between the potentially more restricted and the totipotent stem cells, also draws attention to a question that many scientists in the field have long anticipated and discussed: How does the cellular microenvironment influence or even determine the potency of a stem cell? In other words, are tissue-specific stem cells only specific because they happen to be in a specific tissue? Bjornson et al. have shown that neuronal stem cells can repopulate depleted bone marrow and repopulate the irradiated haematopoietic system just as a blood stem cell would in bone marrow transplantation (Bjornson et al. 1999).

Although the potential of embryonic stem cells – ES and EG cells of the "classical" type or "redifferentiated" versions as in the above-mentioned experiments – is indeed intriguing, their implantation and transplantation are not the only method of how stem cell technology could potentially be used in therapy. Hematopoietic stem cell transplantation has all but replaced "classical" whole bone marrow transplantation. Beyond haematology and oncology though, similar stem cell-based therapies are hampered by the fact that solid organs with their complex structure cannot be easily reconstituted from infused stem cell suspensions. Still, the principle might hold, and this type of stem cell therapy might, as counter-intuitive as this might seem at first, provide a radically new therapeutic approach to cell loss in tissues with low intrinsic regenerative potential. The best and most prominent example of which is the nervous system.

3 Stem Cell Biology and Neurology

Transplantation strategies in neurology have advanced to a clinical stage within the last 20 years. To date, approximately 275 patients with Parkinson's disease have received transplants, some of them in Sweden (Brundin et al. 2000). Although some of these show a very encouraging outcome (Hagell et al. 2000; Wenning et al. 1997), the technique has been under strong opposition because the transplanted brain tissue is derived from the ventral mesencephalon of aborted embryos. It is very likely, but it has not yet been proven, that the cells that mediate the positive therapeutic effect are indeed neuronal stem cells which are transferred within the transplanted tissue. ES or EG cells could be obviously used for this same purpose for Parkinson's and other diseases (Brustle et al. 1999), but similar ethical barriers apply. A fundamental change could, therefore, be achieved if stem cells could be derived from the adult organism, especially the patient himself, so that ultimately an autologous transplantation would be possible. Beyond ethical problems, many practical issues, particularly with regard to immunological and other incompatibilities, could be resolved.

The idea of using adult stem cells is less utopian that it may seem at first because the adult nervous system apparently does indeed harbour neuronal stem cells in great number. At least from the adult rodent brain these stem cells have been successfully brought into cell culture, propagated and studied in detail. From Reynolds' and Weiss' first evidence in 1992 that a multipotent progenitor cell can be derived from the mouse forebrain (Reynolds and Weiss 1992), research has advanced to the recent demonstration that stem cells in the strict sense of the definition can be extracted from brain regions as diverse as hippocampus (Palmer et al. 1997), spinal cord (Shihabuddin et al. 1997), neocortex, septum, striatum and even white matter tracts such as the corpus callosum and the optic nerve (Palmer et al. 1995, 1999). At least some of these cells fulfil the strict criteria to be named stem cells, but to date it remains unresolved how large their potential in vivo is. In a way these stem cells are an culture "artefact", and one could argue that as soon as they hit the culture dish, not very many conclusions about their character in vivo can be drawn. The extreme examples that brain-derived neuronal stem cells could repopulate a depleted mouse bone marrow (Bjornson et al. 1999) or could be "reprogrammed" to become ES-like cells (Clarke et al. 2000) indicate a tremendous potential of these cells; however, it is entirely unknown how much of this potential is only a result of taking the cells out of the brain. For many practical and applied questions, e.g. those around transplantation approaches, this does hardly matter. To understand the fundamental biological properties of these cells, however, it will be important, and with regard to the question why these cells exist in the adult brain and what there inherent function might be, it becomes indispensable to study neuronal stem cells in vivo. One might want to know whether these cells play an important role in neuronal health and disease and whether this knowledge might represent a key to future therapeutic strategies beyond transplantation.

Fortunately, there is a way to study neuronal stem cells in vivo and to learn about their functional properties under "real-life conditions". In two regions of the adult brain, new neurons are continually born from dividing neuronal stem or progenitor cells. One is the subventricular zone of the lateral walls of the lateral ventricles (Corotto et al. 1993), where new cells are generated in large number and migrate along a preformed path into the olfactory bulb (Lois et al. 1996), where they differentiate into neurons of at least two distinct phenotypes (H.G. Kuhn, personal communication). As olfactory receptor neurons in the olfactory epithelium have a high turnover (Calof et al. 1996), one has reasoned that this adult neurogenesis in the olfactory system is necessary to allow sufficient plasticity in the connections between the new receptors and neurons elsewhere in the brain. It has also been shown in rats that closing of one naris results in a decrease in neurogenesis on the same side (Corotto et al. 1994), while reopening stimulates repopulation of the atrophied olfactory bulb through increased neurogenesis (Cummings et al. 1997). There is some evidence that the newly generated neurons in this system are indeed required for odour discrimination, but so far this evidence is correlational (Gheusi et al. 2000). Still, it is obvious that adult neurogenesis is tightly knit into a regulatory network related to the overall function of the system.

The second region in which neurogenesis occurs in the adult brain is the dentate gyrus of the hippocampus (Altman and Das 1965; Cameron et al. 1993; Kaplan and Hinds 1977; Kuhn et al. 1996). Here, the population of dividing stem or progenitor cells is situated on the border between the granule cell layer and the so-called hilus. The cells proliferate, spin off daughter cells that are either dying rather soon or migrate into the granule cell layer and express markers of neuronal maturation. They extend dendrites into the molecular layer and axons along the mossy-fibre tract to region CA3, just as do the other granule cells that were born previously (Markakis and Gage 1999; Stanfield and Trice 1988). This entire process, not only the act of cell proliferation, is called "neurogenesis". This is important because several mechanisms have been identified that increase proliferation of the progenitor cells, without necessarily stimulating net neurogenesis to a similar extent. By strain comparisons in mice we could show that several steps in this regulation are controlled rather independently and that proliferation alone is no sufficient predictor of how many neurons are finally generated (Kempermann et al. 1997a). This implies that any extrinsic attempts to regulate adult neurogenesis will have to take into account that several regulatory steps have to be influenced to obtain a net increase in neurogenesis. While proliferation is the most obvious step and is often erroneously equalled with "neurogenesis", others are survival (indicating a selection process), migration, differentiation, extension of neurites, formation of synapses, generation of action potentials and activity in functional circuits. Some of these steps have been studied in vivo, none however in any greater detail, and the underlying molecular mechanisms are largely unknown.

Since 1965 it has been known that neurogenesis does occur in the adult mammalian brain, but the finding seemed to be so out of line with the prevailing concepts of neurobiology that it had to be rediscovered several times (Altman et al. 1965; Kaplan et al. 1977; Cameron et al. 1993; Kuhn et al. 1996). Finally, in 1998 Ericksson and co-workers could prove in a clever experiment that the adult human hippocampus does generate new neurons even at the age of over 70 years (Eriksson et al. 1998). This result could be considered as a proof of principle: the adult human brain does contain both the cellular source (i.e. the stem cells) and the opportunity (i.e. the regulatory apparatus and micro-environmental conditions) to produce new nerve cells.

Studying adult neurogenesis is a way of studying neuronal stem cells in their "natural environment" in the adult brain. The major obstacle for this research is that, to date, no marker for neuronal stem cells is known. As a consequence, the stem or progenitor cells can only be recognized in situ by labelling them while they undergo division and analysing post hoc into which phenotype they later differentiate. The now commonly used marker for proliferating cells is bromodeoxyuridine (BrdU), a thymidine analog that is incorporated into the newly synthesized DNA during the S-phase of mitosis and that can be detected immunohisto-chemically (del Rio and Soriano 1989). Co-labelling for BrdU and differentiation markers allows one to determine whether a newly born cell has become a neuron (or any other cell type for which a specific marker is known). Unlike haematology, where a number of surface antigens has been identified that allows one to distinguish different types of (stem) cells and describe the cascades of cellular development in quite some detail, the approach that is available to neurobiologists remains rather coarse. We therefore do not know yet how homogeneous the population of neuronal stem or progenitor cells in the dentate gyrus is. Still, even with this lack in precision, a number of interesting questions has been answered.

Adult neurogenesis is the reason why adult neuronal stem cells are interesting. The result defines the entire process. As neuronal stem cells can apparently be found anywhere in the brain (Palmer et al. 1999) neurogenesis depends not only on the neuronal stem cells but also on a microenvironment that is permissive for neurogenesis. One way to study aspects of this permissiveness is to examine the regulation of adult neurogenesis. For example, it has been shown that stress suppresses adult neurogenesis (Gould et al. 1998), a down-regulation that might be largely mediated by glucocorticoids. Removal of endogenous glucocorticoids disinhibits neurogenesis; application of exogenous glucocorticoids reconstitutes the suppression (Cameron and Gould 1994). Similarly, the excitatory neurotransmitter glutamate that transmits the largest input to the dentate gyrus (from the entorhinal cortex) regulates adult neurogenesis. Glutamatergic deafferentation results in an increase in proliferative activity (Cameron et al. 1995). Interestingly, however, a very strong excitatory stimulus, such as in an experimental seizure, has a similar effect (Bengzon et al. 1997; Parent et al. 1997). It is not known quantitatively how this stimulation translates into net neurogenesis, but the thought is intriguing that there might be a level of excitatory input that is optimal for neurogenesis and, thus, that adult neurogenesis is tightly regulated by function. This thought is further substantiated by the finding that exposure to environmental stimuli, that is for the laboratory animals living in a so-called "enriched environment", results in a 60% increase in neurogenesis (Kempermann et al. 1997b, 1998a,b). At the same time, these animals perform better on a learning task, which suggests (but does not prove) that there is a causal link between the two findings. A related finding was that running stimulates adult neurogenesis (Van Praag et al. 1999b), but in contrast to the enriched environment which primarily seemed to promote the survival of newly generated cells, physical activity had a strong effect on the proliferation of cells in the dentate gyrus. Similarly, the running animals did better on a learning task, and it was shown that LTP ("long-term potentiation") is enhanced, the known electrophysiological correlate to learning (Van Praag et al. 1999a). It still remains to be shown that it is indeed the new cells that produced this enhancement, but the correlation is suggestive. A brief learning stimulus alone also caused an increased survival of newborn cells (Gould et al. 1999), strengthening the hypothesis that the activity-dependent regulation of neurogenesis is an intrinsic property of the adult hippocampus. It is not known which factors mediate the activity-dependent regulation of adult neurogenesis; but beyond excitatory neurotransmitters and glucocorticoids, serotonin expression has been shown to positively correlate with adult hippocampal neurogenesis (Brezun and Daszuta 1999, 2000). Serotonergic fibres project from the reticular formation to the dentate gyrus and might thus be involved in those regulatory effects linked to physical activity. Sexual hormones might be involved in the baseline regulation of neurogenesis, but there does not seem to be a strong sex difference (Tanapat et al. 1999). Many more possible regulators remain to be studied, but even now it is obvious that regulation of neurogenesis is complex and that it will be hard to distinguish primary and secondary effects.

4 Conclusion

The finding that adult neurogenesis is regulated in an activity-dependent manner suggests that neurogenesis, although it is locally restricted, is a physiological property of the adult brain and not so much an exception, as was previously thought. This implies that the underlying mechanisms are not out of reach for approaches that intend to make use of these mechanisms for stem cell-based therapies. Ultimately, the same rules of neurogenic permissiveness will apply to endogenous stem cells involved in neurogenesis in situ, and to transplanted embryonic or adult stem cells. Therefore, research on neurogenic permissiveness lies right at the heart of neuronal stem cell biology.

References

Altman J, Das GD (1965) Autoradiographic and histologic evidence of postnatal neurogenesis in rats. J Comp Neurol 124:319–335

Bengzon J, Kokaia Z, Elmér E et al (1997) Apoptosis and proliferation of dentate gyrus neurons after single and intermittent limbic seizures. Proc Natl Acad Sci USA 94:10432–10437

Bjornson CRR, Rietze RL, Reynolds BA et al (1999) Turning brain into blood: a hematopoietic fate adopted by adult neural stem cells in vivo. Science 283:534–537

Brezun JM, Daszuta A (1999) Depletion in serotonin decreases neurogenesis in the dentate gyrus and the subventricular zone of adult rats. Neuroscience 89:999–1002

Brezun JM, Daszuta A (2000) Serotonin may stimulate granule cell proliferation in the adult hippocampus, as observed in rats grafted with foetal raphe neurons. Eur J Neurosci 12:391–396

Brundin P, Pogarell O, Hagell P et al (2000) Bilateral caudate and putamen grafts of embryonic mesencephalic tissue treated with lazaroids in Parkinson's disease. Brain 123:1380–1390

Brüstle O, Choudhary K, Karram K et al (1998) Chimeric brains generated by intraventricular transplantation of fetal human brain cells into embryonic rats. Nat Biotech 16:1040–1044

Brustle O, Jones KN, Learish RD et al (1999) Embryonic stem cell-derived glial precursors: a source of myelinating transplants. Science 285:754–756

Calof AL, Hagiwara N, Holcomb JD et al (1996) Neurogenesis and cell death in olfactory epithelium. Journal of Neurobiology 30:67–81

Cameron HA, Gould E (1994) Adult neurogenesis is regulated by adrenal steroids in the dentate gyrus. Neuroscience 61:203–209

Cameron HA, McEwen BS, Gould E (1995) Regulation of adult neurogenesis by excitatory input and NMDA receptor activation in the dentate gyrus. J Neurosci 15:4687–4692

Cameron HA, Woolley CS, McEwen BS et al (1993) Differentiation of newly born neurons and glia in the dentate gyrus of the adult rat. Neuroscience 56:337–344

Clarke DL, Johansson CB, Wilbertz J et al (2000) Generalized potential of adult neural stem cells. Science 288:1660–1663

Corotto FS, Henegar JA, Maruniak JA (1993) Neurogenesis persists in the subependymal layer of the adult mouse brain. Neurosci Lett 149:111–114

Corotto FS, Henegar JR, Maruniak JA (1994) Odor deprivation leads to reduced neurogenesis and reduced neuronal survival in the olfactory bulb of the adult mouse. Neuroscience 61:739–744

Cummings DM, Henning HE, Brunjes PC (1997) Olfactory bulb recovery after early sensory deprivation. J Neurosci 17:7433–7440

Donovan PJ (1998) The germ cell – the mother of all stem cells. Int J Dev Biol 42:1043–1050

Eriksson PS, Perfilieva E, Björk-Eriksson T et al (1998) Neurogenesis in the adult human hippocampus. Nat Med 4:1313–1317

Gage F (2000) Mammalian neural stem cells. Science 287:1433–1438

Gheusi G, Cremer H, McLean H et al (2000) Importance of newly generated neurons in the adult olfactory bulb for odor discrimination. Proc Natl Acad Sci USA 97:1823–1828

Gould E, Beylin A, Tanapat P et al (1999) Learning enhances adult neurogenesis in the hippoampal formation. Nat Neurosci 2:260–265

Gould E, Tanapat P, McEwen BS et al (1998) Proliferation of granule cell precursors in the dentate gyrus of adult monkeys is diminished by stress. Proc Natl Acad Sci USA 95:3168–3171

Hagell P, Crabb L, Pogarell O et al (2000) Health-related quality of life following bilateral intrastriatal transplantation in Parkinson's disease. Mov Disord 15:224–229

Kaplan MS, Hinds JW (1977) Neurogenesis in the adult rat: electron microscopic analysis of light radioautographs. Science 197:1092–1094

Kempermann G, Brandon EP, Gage FH (1998a) Environmental stimulation of 129/SvJ mice results in increased cell proliferation and neurogenesis in the adult dentate gyrus. Curr Biol 8:939–942

Kempermann G, Kuhn HG, Gage FH (1997a) Genetic influence on neurogenesis in the dentate gyrus of adult mice. Proc Natl Acad Sci USA 94:10409–10414

Kempermann G, Kuhn HG, Gage FH (1997b) More hippocampal neurons in adult mice living in an enriched environment. Nature 386:493–495

Kempermann G, Kuhn HG, Gage FH (1998b) Experience-induced neurogenesis in the senescent dentate gyrus. J Neurosci 18:3206–3212

Kuhn HG, Dickinson-Anson H, Gage FH (1996) Neurogenesis in the dentate gyrus of the adult rat: age-related decrease of neuronal progenitor proliferation. J Neurosci 16:2027–2033

Lois C, Garcia-Verdugo J-M, Alvarez-Buylla A (1996) Chain migration of neuronal precursors. Science 271:978–981

Markakis E, Gage FH (1999) Adult-generated neurons in the dentate gyrus send axonal projections to the field CA3 and are surrounded by synaptic vesicles. J Comp Neurol 406:449–460

Palmer TD, Markakis EA, Willhoite AR et al (1999) Fibroblast Growth Factor-2 activates a latent neurogenic program in neural stem cells from divers regions of the adult CNS. J Neurosci 19:8487–8497

Palmer TD, Ray J, Gage FH (1995) FGF-2-responsive neuronal progenitors reside in proliferative and quiescent regions of the adult rodent brain. Mol Cell Neurosci 6:474–486

Palmer TD, Takahashi J, Gage FH (1997) The adult rat hippocampus contains premordial neural stem cells. Mol Cell Neurosci 8:389–404

Parent JM, Yu TW, Leibowitz RT et al (1997) Dentate granule cell neurogenesis is increased by seizures and contributes to aberrant network reorganization in the adult rat hippocampus. J Neurosci 17:3727–3738

Reynolds BA, Weiss S (1992) Generation of neurons and astrocytes from isolated cells of the adult mammalian central nervous system [see comments]. Science 255:1707–1710

Rio JA del, Soriano E (1989) Immunocytochemical detection of 5¢-bromodeoxyuridine incorporation in the central nervous system of the mouse. Dev Brain Res 49:311–317

Shamblott MJ, Axelman J, Wang S et al (1998) Derivation of pluripotent stem cells from cultured human premordial germ cells. Proc Natl Acad Sci USA 95:13726–13731

Shihabuddin LS, Ray J, Gage FH (1997) FGF-2 is sufficient to isolate progenitors found in the adult mammalian spinal cord. Exp Neurol 148:577–586

Stanfield BB, Trice JE (1988) Evidence that granule cells generated in the dentate gyrus of adult rats extend axonal projections. Exp Brain Res 72:399–406

Tanapat P, Hastings NB, Reeves AJ et al (1999) Estrogen stimulates a transient increase in the number of new neurons in the dentate gyrus of the adult female rat. J Neurosci 19:5792–5801

Thomson J, Odorico J (2000) Human embryonic stem cell and embryonic germ cell lines. Trends Biotechnol 18:53–57

Van Praag H, Christie BR, Sejnowski TJ et al (1999a) Running enhances neurogenesis, learning and long-term potentiation in mice. Proc Natl Acad Sci USA 96:13427–13431

Van Praag H, Kempermann G, Gage FH (1999b) Running increases cell proliferation and neurogenesis in the adult mouse dentate gyrus. Nat Neurosci 2:266–270

Wenning GK, Odin P, Morrish P et al (1997) Short- and long-term survival and function of unilateral intrastriatal dopaminergic grafts in Parkinson's disease. Ann Neurol 42:95–107

3 Tissue Engineering by Cell Transplantation

P.V. Shastri, I. Martin

1 Historical Perspective

It has long been known that dissociated mammalian cells are capable of forming sheets of tissue in monolayer cultures (Steinberg 1963). The recognition that the final form of a mass of cells can be influenced and dictated by associating it with a scaffolding material led to the emergence of Tissue Engineering (TE). The earliest attempts at engineering a tissue mass using the principles of TE were carried out by Bell et al. (1979, 1981a,b) and Yannas et al. (1980, 1982; Yannas and Burke 1980) at the Massachusetts Institute of Technology in the late 1970s early 1980s. Their approach relied on the use of collagen-based gels and

Presented at the 35th Workshop sponsored by the Ernst Schering Research Foundation on "Stem Cell Transplantation and Tissue Engineering", 3-5 August 2000, Hannover, Germany foams to provide the necessary structural definition for the proliferation and differentiation of neonatal human foreskin fibroblasts into an epidermis-like tissue. While these experiments proved the feasibility of engineering a viable, well-defined mass of tissue with biological functionality, the use of a collagen-based scaffold displayed some serious drawbacks. The shrinkage of the scaffold under the contractile forces exerted by the cells and the immunological issues associated with the use of bovine collagen are still of primary concern. Furthermore, due to the difficulty of processing collagen into large, complex structures, TE using collagen has been restricted to membranes or sheets of tissue, mostly suited to the engineering of skin equivalents.

2 Coming of Age: Synthetic Polymer Scaffolds

In the mid 1980s, Langer and Vacanti proposed that a highly porous synthetic biodegradable polymer structure could serve as a scaffold for TE and provide certain unique advantages over the biological-derived scaffolds, particularly with respect to fashioning complex and well-defined three-dimensional environments and alleviating the immunological concerns (Vacanti et al. 1988, 1991). Synthetic scaffolds also offer a whole host of other possibilities with respect to tailoring important scaffold properties such as degradation rate, mechanical properties, geometry, surface chemistry, and biological functionality (Vacanti et al. 1988; Langer and Vacanti 1993; Vacanti and Langer 1999). Good mechanical strength is an important consideration in engineering of load bearing tissues such as cartilage and bone. The approach proposed by Langer and Vacanti induced a paradigm shift, and has ushered in a new thinking and numerous new possibilities for TE. Thus, for the first time it was possible to predictably design and culture tissues of complex shapes. This is exemplified by the grafting of a polymer construct seeded with chondrocytes in a nude mouse to generate a cartilaginous tissue in the shape of a human ear (Gao et al. 1998).

3 Innovative Polymer Processing Techniques

The polymer scaffold used in engineering the ear on the back of a mouse was a fibrous felt of poly(glycolic acid) (PGA), a degradable polymer derived from glycolic acid. The fibrous nature of this scaffold does not allow fabrication of structures of complex shapes and with good mechanical properties. As a consequence, new approaches such as fiber bonding and solvent casting/particulate salt leaching (Mikos et al. 1993, 1994) have been developed to produce mechanically stronger scaffolds. While these methods improve the mechanical characteristics of the foam, they are severely deficient with respect to control over the porosity and pore characteristics. They are also limited in their applicability and cannot yield porous structures greater than a few millimeters (mm) in thickness (Lu and Mikos 1996). High porosity with an interconnected pore structure is important for the generation of a viable, uniform mass of tissue. The search for a process that would yield scaffolds with high porosities and good mechanical properties with no thickness limitation has been on ever since.

Recently, a novel process to produce porous polymeric matrices with enhanced control over pore structure, porosity, mechanical properties, degradation rate, and chemical composition has been developed (Shastri et al. 2000). This process is applicable to a large family of polymers and more importantly is capable of yielding large structures in excess of a few inches in thickness with excellent handling characteristics. It yields porous polymeric structures with a continuous polymer phase and smooth, highly open-pore morphology. The presence of a continuous tubular polymer phase also allows for a more uniform distribution of stress through the scaffold, which therefore makes it less likely to buckle. A continuous polymer structure, when compared to a fibrous one, has been shown to be advantageous in the culturing of pluripotent mesenchymal cells, bone marrow stromal cells (BMSC), in vitro. Under identical culture conditions, PGA fibrous mesh scaffolds collapsed and reduced to a volume that was less than 25% of the original volume within 3 days, while those possessing a continuous polymer structure retained their geometry for over 4 weeks (Fig. 1) (Martin et al. 2001). The highly porous and mechanically robust environment of the polymer scaffold translated into a more homogeneous distribution of cells and extracellular matrix (ECM) components (Martin et al. 2001). Complex

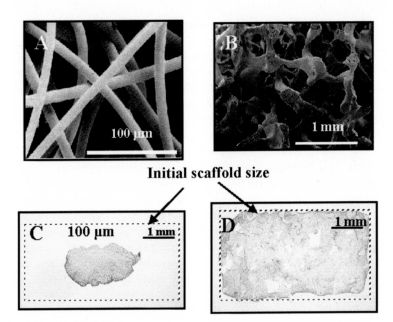

Fig. 1A–D. Influence of scaffold mechanical characteristics on bone marrow stromal cells (BMSC) proliferation. **A** A non-woven mesh derived of poly(gly-colic acid) (PGA). **B** Foam possessing a continuous polymer phase derived from blending poly(L-lactic acid-co-glycolic acid) with poly(ethylene glycol) in 80:20 w/w ratio (PLGA/PEG); **C, D** Structural fate of the scaffold seeded with BMSC after 4 weeks culture, haematoxylin–eosin-stained cross-sections. Note that the non-woven PGA mesh (**C**) has collapsed to less than 40% of its original dimensions, causing exclusion of the cells to the periphery of the scaf-fold (*intense pink coloration along the edges*) while the PLGA/PEG foam has retained its original dimensions, thus promoting a uniform distribution of cells (*uniform pink coloration*) through the scaffold

cartilaginous structures such as one in the likeness of a human nose have been fashioned in vitro using a scaffold produced by the above-men-tioned process (Shastri et al. 2000).

Surgical medicine is slowly but surely moving towards minimally invasive procedures. Therefore, materials that can be shaped and formed in situ will have utility in developing minimally invasive modalities for cell transplantation and tissue reconstruction. Minimally invasive tech-

nology has particular relevance in facial and musculo-skeletal reconstruction. Effort towards this goal is already underway. Biodegradable
polymeric systems that can be shaped using light radiation have been
developed and successfully evaluated in animal models for bone and
cartilage reconstruction (Anseth et al. 1999; Elisseeff et al. 2000).

Polymer structure and its chemical make-up play an important role in
the interaction of the cells with the scaffold and the subsequent evolution of the tissue. Pioneering studies carried out by Folkman and
Moscona have unequivocally shown that the chemical nature of the
polymer dictates cell behavior on the surface and that the biological
function of the cell is closely related to its shape (Folkman and Moscona
1978). This has been well established for various cell types, including
chondrocytes and hepatocytes (Ben-Zeev et al. 1988; Dunn et al. 1989).
Therefore, the chemical composition and structure of polymer scaffolds
have to be designed in a manner to induce the appropriate biological
outcome in a desired cell type. Several approaches have been taken to
improve the specificity of scaffold surfaces towards cells. These include
chemical surface modification to render the surface more hydrophilic
(Gao et al. 1990), covalent binding of peptide sequences such as RGD,
IKVAV, and YIGSR (Cook et al. 1997), and tethering of growth factors
(Kuhl and Griffith-Cima 1996, 1997). However, the lack of simplicity of
these approaches diminishes their clinical feasibility. Hence, simpler
methodologies need to be developed. One such approach to improve
cell–scaffold interactions, which may be simple and versatile, is based
on true physical blends of synthetic and biological polymers. Such
blended substrates may possess the correct surface chemistry for chemi-
sorption and delivery of growth factors while providing improved
cell–substrate interactions (Shastri et al. 2000). Alternatively, semi-synthetic resorbable materials can be developed starting from naturally
occurring polymers. An example of such a semi-synthetic biomaterial
that holds promise for TE applications is based on alkyl esters of
purified hyaluronan (Campoccia et al. 1998). By varying the chemical
make up of the alkyl side chain and the extent of esterification, the
biological properties of the hyaluronan can be considerably affected to
obtain materials that completely favor or inhibit the adhesion of specific
cell types (Campoccia et al. 1998).

4 The Holy Grail of Tissue Engineering: Stem Cells

The generation of specific tissues using the principles of TE is very often limited by (1) shortage of donor cells and/or (2) poor proliferative potential of terminally differentiated cells. These issues could be effectively solved using stem cells (Brustle and McKay 1996; Asahara et al. 1997; Thomson et al. 1998; Bianco and Robey 2000). Stem cells have an indefinite self-renewal potential and are capable of undergoing selective differentiation into multiple distinct cell lineages according to specific culture conditions. The advantages of stem cells over mature, differentiated cells include: (1) high proliferative capacity, (2) high migratory capacity, (3) ability to differentiate into multiple cell types, and (4) high responsiveness to growth factors and signaling molecules (Thomson et al. 1998; Bianco and Robey 2000; Martin et al. 1997). In recent years, a tremendous effort has gone into the identification and isolation of mammalian stem cells. As a result, stem cell populations that are precursors to the musculo-skeletal system (bone marrow stem cells, satellite cells) (Bianco and Robey 2000), the vasculature (angioblasts, hemangioblasts, hematopoietic cells) (Asahara et al. 1997; Shi et al. 1998; Rafii 2000; Lin et al. 2000), hepatic system (Petersen et al. 1999), and neuronal tissue (neural stem cells) (McKay 1997; Cage 2000; Momma et al. 2000; Clarke et al. 2000) have been discovered. Currently, several studies are underway to engineer functional replacement tissues, particularly in the neuronal (McDonald et al. 1999) and musculo-skeletal areas (Taylor et al. 1998), using stem cells. The challenges in using stem cells for tissue engineering are the ability (1) to expand cells efficiently while preventing a spontaneous non-specific differentiation; and (2) to provide the needed stimuli for cell differentiation towards the lineage of interest. Perpetuating stem cells in vitro is typically approached by using a combination of epigenetic and genetic procedures (Villa et al. 2000) and/or by culturing cells in a medium supplemented with specific growth factors (Martin et al. 1997; Villa et al. 2000). In order to achieve a selective differentiation of stem cells, several factors of a different nature have to be taken into account. These include biochemical regulatory molecules, structural and chemical characteristics of the scaffold, and application of mechanical forces or electromagnetic fields.

Our own experience in using BMSC to engineer large segments of bone-like and cartilaginous tissues using the principles of tissue engi-

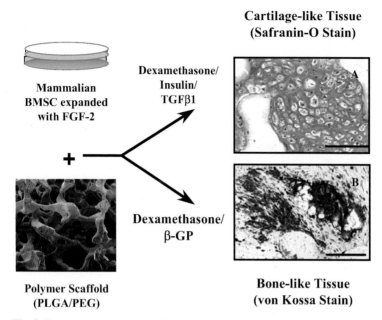

Fig. 2. Bone marrow stromal cells (*BMSC*) expanded in vitro in presence of fibroblast growth factor-2 (*FGF-2*) and associated with a mechanically stable polymer scaffold can be differentiated in vitro into either a cartilaginous (*top right*) or bone-like (*bottom right*) tissue upon supplementation with the appropriate soluble differentiation promoting factors. Safranin-0 stains the glycosaminoglycan (GAG)-rich regions in the tissue while the von Kossa silver stain stains the mineralized regions in the tissue. *Scale bar*=100 μm. *TGFβ1*, transforming growth factor beta-1; β-*GP*, beta glycerol phosphate

neering have shown that it is possible to predictably commit the same population of BMSC to either the osteogenic or the chondrogenic phenotype by combining soluble signaling molecules with specific scaffold structural characteristics (Fig. 2) (Martin et al. 2001). We have also shown that non-invasive approaches such as electromagnetic (EM) stimulation via a conductive polymer substrate may be useful in increasing the osteogenic potential of BMSC (Shastri et al. 1999). Such application of EM stimulation was previously shown to enhance the differentiation of PC-12 cells in vitro as ascertained by the increase in the length of neurite processes (Shastri et al. 1996; Schmidt et al. 1997) and may prove useful in guided tissue regeneration of the central and peripheral nervous system (Shastri and Pishoko 1998; Shastri et al. 1996). Transplantation studies into telencephalic vesicles of embryonic rats using neural precursor derived from dorsal and ventral mouse forebrain have shown that neuronal migration and differentiation are predominantly regulated by non-cell-autonomous signals (Brustle et al. 1995). This suggests that structural guidance is important in restoring neuronal function.

Specific regulatory molecules can be provided to cells not only by their supplementation to the culture medium, but also by localized delivery using controlled release technologies (Langer and Vacanti 1993; Langer 1998) and genetic engineering of cells (Wilson et al. 1988, 1989; Mulligan 1993; Levy et al. 1998; Gussoni et al. 1999; Gojo et al. 2000). Of the three approaches, transfection of cells to induce secretion of growth factors and/or deficient genes (Springer et al. 1998; Powell et al. 1999) is likely to have the greatest potential for long-term impact. However, before this approach can become clinically acceptable, a clear understanding of the consequences of upregulation of mRNA expression in one cell system on a plural environment and the overall physiology has to be established. Several studies exploring the potential of gene therapy using neural, endothelial, and mesenchymal progenitor cells are currently underway with promising results (Gussoni et al. 1999; Hickman et al. 1994; Ksner 1998; Asahara et al. 2000; Benedetti et al. 2000; Han and Fischer 2000).

It is well established that vascularization is key in the development of an embryo (Baldwin 1996). It is essential for defining the environment around a dividing cell and to localize and transport signaling molecules in most developing tissues. The process of vasculogenesis is followed by neurogenesis, which ensures "wiring" of the growing tissues to the central nervous system. Vascularization is very important to maintain tissue viability and allow for transport of nutrients and endogenous growth factors to the site of the transplanted tissue; while innervation is essential for tissues such as muscle where the function is dictated by sensory perceptions. Therefore, most engineered tissues should be vascularized and innervated before they become fully functional in vivo. Neural and endothelial progenitor cells can be co- and pre-seeded onto scaffolds to create environments that are supportive for innervation and vascularization of the engineered tissue. Such composite tissues can also serve as model systems to study interactions between different cell types and the effect of gene upregulation at a cellular/local level on the evolution of distinct tissues. This understanding is also essential to ensure the best clinical outcome in disease treatment using gene therapy.

5 Cell Seeding and Cell–Polymer Construct Cultivation

Efficient seeding of cells onto porous scaffolds is an important prerequisite for the development of a homogeneous tissue from the lowest possible cell number. Static seeding by simple cell loading is generally associated with low efficiencies and inhomogeneous cell distributions. Dynamic seeding using spinner flasks can achieve higher efficiencies of seeding, approaching 100% (Vunjak-Novakovic et al. 1998; Burg et al. 2000), but the cell density is generally higher at the scaffold periphery and the process exposes cells and constructs to turbulent fluid flow. Alternative approaches include delivery of cells from the scaffold center outwards (Wald et al. 1993), suspension of cells into a gel material used as a void filler (Sittinger et al. 1994), application of vacuum (van Wachen et al. 1990), and electrostatic seeding (Bowlin and Rittgers 1997). The choice of the seeding procedure should be primarily dictated by the chemistry of the scaffolding material and the cell type, but up to the present time it has not yet been clearly defined which cell-seeding procedure is more suitable for which specific scaffold/cell system.

Once the cells are associated to the scaffold, cell–polymer constructs can be cultured in appropriate bioreactors, i.e., in devices providing a controlled environment for specific physical parameters. Several bioreactors have been developed, which allow application of specific regimes of fluid flow, shear stresses, tension, or compression to developing tissues. Based on recent studies where bioreactors were used to improve the generation of cartilage (Freed et al. 1998), bone-like structures (Qiu et al. 1999), tendons (Hsieh et al. 2000), blood vessels (Niklason et al. 1999), and skeletal muscles (Kim et al. 1998), it is clear that the future of tissue engineering will be highly dependent on the development of suitable bioreactors for the culture of cell-polymer constructs under appropriate conditions.

6 Tissue Engineering in Drug Delivery

Current cell-based therapies for localized or systemic drug delivery are focused on the encapsulation and delivery of terminally differentiated cells (for example, islet cells or dopamine-secreting neurons) using complex, implantable, synthetic polymer-derived devices. This poses several problems namely: (1) sustained immune response to the delivery device, (2) poor transport of the therapeutic agent due to encapsulation of the delivery device in a fibrous capsule, (3) long-term decrease in cell viability, and (4) long-term side effects of the implantable device. The principles of TE, combined with those of cell transfection, offer numerous hitherto unavailable avenues to exploit cell-based drug delivery therapies. As an example, autologous human skeletal myoblasts from a potential patient population have been isolated, genetically modified to secrete foreign proteins, and tissue engineered into implantable living protein secretory devices for therapeutic use (Powell et al. 1999). In another study, experimental brain tumors have been treated by grafting neural stem cells engineered to produce therapeutic molecules (Benedetti et al. 2000). In the coming years, it would be tempting to envision drug delivery vehicles that are entirely cellular in origin, with multiple therapeutic reservoirs within the same core, and a responsive barrier that is controlled by the patient's physiology, all derived from the patient's own cells (Fig. 3). Under the appropriate conditions, this transplanted tissue, which in essence is a bioactive drug delivery device, could become integrated in the recipient's physiology in a manner to become self-regulated. So far, regulation in a delivery system has been achieved only using external devices and triggers such as pumps. A responsive-controlled release system, that is based on the biological processes of the recipient and not on any external trigger, would be a quantum leap in the area of controlled and responsive delivery systems and a boon to patients who have to be under complex regimen of sustained medication.

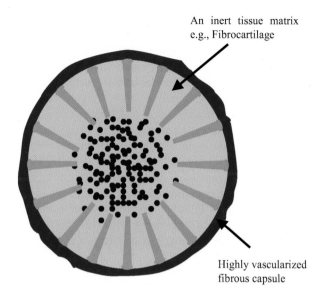

An inert tissue matrix
e.g., Fibrocartilage

Highly vascularized
fibrous capsule

● Cells genetically engineered to produce a desired bioactive molecule

▌ Channels lined with smooth muscle cells

Fig. 3. An example of a completely cellular drug delivery system. In this example, cells that are genetically engineered to secrete the biomolecule of interest (*blue solid circles*) are grown within a cavity of a biologically inert tissue such as fibrocartilage. This inert fibrocartilage matrix is interspersed with radial channels originating from the core, derived from cell types that are capable of undergoing contraction and relaxation upon exposure to biochemical or electrical signals, such as smooth muscle cells. The purpose of these channels is to control the diffusion of the biomolecule from the reservoir (mass of genetically engineered cells). The outer layer of this delivery device is composed of vascularized fibrous tissue. The purpose of this outer layer is twofold: (1) to prevent infiltration of the surrounding tissue into the implant and (2) to act as a highly permeable barrier

7 Future of Tissue Engineering

TE is at the crossroads. During the past two decades, TE has evolved from being a craft into a highly interdisciplinary field that encompasses chemistry, cellular and molecular biology, materials science, and chemical engineering. It has matured into a highly challenging scientific endeavor with immense clinical and commercial potential. Numerous patents issued in this field are a testimony to this transformation (Pabst 1999). Several tissues including skin (Bell et al. 1991), cartilage (Vacanti et al. 1991; Cao et al. 1997), blood vessels (Niklason et al. 1999), and bladders (Oberpenning et al. 1999) have been created using principles of TE and are currently in various stages of development. While the current emphasis of TE is the generation of viable, functional tissues for transplantation, one can envisage a future where TE will become a powerful tool in harnessing and delivering the benefits of gene-based medicine.

References

Anseth K, Shastri V, Langer R (1999) Photopolymerizable degradable polyanhydrides with osteocompatibility. Nat Biotechnol 17(2):156–159

Asahara T, Murohara T, Sullivan A, Silver M, Zee R van der, Li T, Witzenbichler B, Schatteman G, Isner JM (1997) Isolation of putative progenitor endothelial cells for angiogenesis. Science 275:964–967

Asahara T, Kalka C, Isner J (2000) Stem cell therapy and gene transfer for regeneration. Gene Ther 7(6):451–457

Baldwin H (1996) Early embryonic vascular development. Cardiovasc Res 31:E34–45

Bell E, Ivarsson B, Merrill C (1979) Production of a tissue-like structure by contraction of collagen lattices by human fibroblasts of different proliferative potential in vitro. Proc Natl Acad Sci USA 76(3):1274–1278

Bell E, Ehrlich H, Sher S, Merrill C, Sarber R, Hull B, Nakatsuji T, Church D, Buttle D (1981a) Development and use of a living skin equivalent. Plast Reconstr Surg 67(3):386–392

Bell E, Ehrlich H, Buttle D, Nakatsuji T (1981b) Living tissue formed in vitro and accepted as skin-equivalent tissue of full thickness. Science 211:1052–1054

Bell E, Rosenberg M, Kemp P, Gay R, Green G, Muthukumaran N, Nolte C (1991) Recipes for reconstituting skin. J Biomech Eng 113(2):113–119

Benedetti S, Pirola B, Pollo B, Magrassi L, Bruzzone MG, Rigamonti D, Galli R, Selleri S, Di Meco F, De Fraja C, Vescovi A, Cattaneo E, Finocchiaro G (2000) Gene therapy of experimental brain tumors using neural progenitor cells. Nat Med 6(4):447–450

Ben-Zeev, A, Robinson G, Bucher N, Farmer S (1988) Cell-cell and cell-matrix interactions differentially regulate the expression of hepatic and cytoskeletal genes in primary cultures of rat hepatocytes. Proc Natl Acad Sci USA 85:2161–2165

Bianco P, Robey P (2000) Marrow stromal stem cells. J Clin Invest 105(12):1663–1668

Bowlin G, Rittgers S (1997) Electrostatic endothelial cell seeding technique for small-diameter (<6 mm) vascular prostheses: feasibility testing. Cell Transplant 6(6):623–629

Brustle O, McKay R (1996) Neuronal progenitors as tools for cell replacement in the nervous system. Curr Opin Neurobiol 6(5):688–695

Brustle O, Maskos U, McKay R (1995) Host-guided migration allows targeted introduction of neurons into the embryonic brain. Neuron 15(6):1275–1285

Burg K, Holder WJ, Culberson C, Beiler R, Greene K, Loebsack A, Roland W, Eiselt P, Mooney D, Halberstadt C (2000) Comparative study of seeding methods for three-dimensional polymeric scaffolds. J Biomed Mater Res 51(4):642–649

Cage F (2000) Mammalian neural stem cells. Science 287:1433–1438

Campoccia D, Doherty P, Radice M, Brun P, Abatangelo G, Williams D (1998) Semisynthetic resorbable materials from hyaluronan esterification. Biomaterials 19(23):2101–2127

Cao Y, Vacanti J, Paige K, Upton J, Vacanti C (1997) Transplantation of chondrocytes utilizing a polymer-cell construct to produce tissue-engineered cartilage in the shape of a human ear. Plast Reconstr Surg 100(2):297–302

Clarke DL, Johansson CB, Wilbertz J, Veress B, Nilsson E, Karlstrom H, Lendahl U, Frisen J (2000) Generalized potential of adult neural stem cells. Science. 288:1660–1663

Cook A, Hrkach J, Gao N, Johnson I, Pajvani U, Cannizzaro S, Langer R (1997) Characterization and development of RGD-peptide-modified poly(lactic acid-co-lysine) as an interactive, resorbable biomaterial. J Biomed Mater Res 35(4):513–523

Dunn JCY, Yarmush MC, Koebe HG, Tompkins RG (1989) Hepatocyte function and extracellular matrix geometry: long-term culture in a sandwich configuration. FASEB J 3:174–177

Elisseeff J, McIntosh W, Anseth K, Riley S, Ragan P, Langer R (2000) Photoencapsulation of chondrocytes in poly(ethylene oxide)-based semi-interpenetrating networks. J Biomed Mater Res 51 (2):164–171

Folkman J, Moscona A (1978) Role of cell shape in growth control. Nature 273:345–349

Freed L, Hollander A, Martin I, Barry J, Langer R, Vunjak-Novakovic G (1998) Chondrogenesis in a cell-polymer-bioreactor system. Exp Cell Res 240(1):58–65

Gao J, Niklason L, Zhao X, Langer R (1998) Surface modification of polyanhydride microspheres. J Pharm Sci 87(2):246–248

Gojo S, Cooper D, Iacomini J, LeGuern C (2000) Gene therapy and transplantation. Transplantation 69(10):1995–1999

Gussoni E, Soneoka Y, Strickland C, Buzney E, Khan M, Flint A, Kunkel L, Mulligan R (1999) Dystrophin expression in the mdx mouse restored by stem cell transplantation. Nature 401:390–394

Han S, Fischer I (2000) Neural stem cells and gene therapy: prospects for repairing the injured spinal cord. JAMA 283(17):2300–2301

Hickman M, Malone R, Lehmann-Bruinsma K, Sih T, Knoell D, Szoka F, Walzem R, Carlson D, Powell J (1994) Gene expression following direct injection of DNA into liver. Hum Gene Ther 5(12):1477

Hsieh A, Tsai C, Ma Q, Lin T, Banes A, Villarreal F, Akeson W, Sung K (2000) Time-dependent increases in type-III collagen gene expression in medical collateral ligament fibroblasts under cyclic strains. J Orthop Res 18(2):220–227

Isner J (1998) Arterial gene transfer of naked DNA for therapeutic angiogenesis: early clinical results. Adv Drug Deliv Rev 30(1–3):185–197

Kim B, Putnam A, Kulik T, Mooney D (1998) Optimizing seeding and culture methods to engineer smooth muscle tissue on biodegradable polymer matrices. Biotechnol Bioeng 57(1):46–54

Kuhl P, Griffith-Cima L (1996) Tethered epidermal growth factor as a paradigm for growth factor-induced stimulation from the solid phase. Nat Med 2(9):1022–1027

Kuhl P, Griffith-Cima L (1997) Erratum to paper in 1996. Nat Med 3(1):93

Langer R (1998) Drug delivery and targeting. Nature 392 [Suppl]:5–10

Langer R, Vacanti JP (1993) Tissue engineering. Science 260:920–926

Levy R, Goldstein S, Bonadio J (1998) Gene therapy for tissue repair and regeneration. Adv Drug Deliv Rev 33(1–2):53–69

Lin Y, Weisdorf D, Solovey A, Hebbel R (2000) Origings of circulating endothelial cells and endothelial outgrowth from blood. J Clin Invest 105(1):71–77

Lu L, Mikos A (1996) The importance of new processing techniques in tissue engineering. MRS Bull 21(11):28–32

Martin I, Muraglia A, Campanile G, Cancedda R, Quarto R (1997) Fibroblast growth factor-2 supports ex vivo expansion and maintenance of osteogenic precursors from human bone marrow. Endocrinology 138(10):4456–4462

Martin I, Shastri V, Padera R, Langer R, Yang J, MacJay A, Vunjak-Novakovic G, Freed L (2001) Selective *in vitro* differentiation of mammalian mesenchymal frogenitor cells into three dimensions skeletal tissues. J Biomed Mat Res 55:229–235

McDonald JW, Liu XZ, Qu Y, Liu S, Mickey SK, Turetsky D, Gottlieb DI, Choi DW (1999) Transplanted embryonic stem cells survive, differentiate and promote recovery in injured rat spinal cord. Nat Med 5(12):1410–1412

McKay R (1997) Stem cells in the central nervous system. Science 276(5309):66–71

Mikos A, Bao Y, Cima L, Ingber D, Vacanti J, Langer R (1993) Preparation of poly(glycolic acid) bonded fiber structures for cell attachment and transplantation. J Biomed Mater Res 27(2):183–189

Mikos AG, Thorsen AJ, Czerwonka LA, Bao Y, Langer R (1994) Preparation and Characterization of Poly(L-lactic acid) Foams. Polymer 35:1068–1077

Momma S, Johansson C, Frisen J (2000) Get to know your stem cells. Curr Opin Neurobiol 10:45–49

Mulligan R (1993) The basic science of gene therapy. Science 260:926–932

Niklason L, Gao J, Abbott W, Hirschi K, Houser S, Marini R, Langer R (1999) Functional arteries grown in vitro. Science 284:489–493

Oberpenning F, Meng J, Yoo J, Atala A (1999) De novo reconstitution of a functional mammalian urinary bladder by tissue engineering. Nat Biotechnol 17(2):149–155

Pabst P (1999) Gene therapy and tissue engineering patents abound. Tissue Eng 5(1):79

Petersen BE, Bowen WC, Patrene KD, Mars WM, Sullivan AK, Murase N, Boggs SS, Greenberger JS, Goff JP (1999) Bone marrow as a potential source of hepatic oval cells. Science 284:1168–1170

Powell C, Shansky J, Del Tatto M, Forman D, Hennessey J, Sullivan K, Zielinski B, Vandenburgh H (1999) Tissue-engineered human bioartificial muscles expressing a foreign recombinant protein for gene therapy. Hum Gene Ther 10(4):565–577

Qiu Q, Ducheyne P, Ayyaswamy P (1999) Fabrication, characterization and evaluation of bioceramic hollow microspheres used as microcarriers for 3-D bone tissue formation in rotating bioreactors. Biomaterials 20(11):989–1001

Rafii S (2000) Circulating endothelial precursors: mystery, reality, and promise. J Clin Invest 105(1):17–19

Schmidt C, Shastri V, Vacanti J, Langer R (1997) Stimulation of neurite outgrowth using an electrically conducting polymer. Proc Natl Acad Sci USA 94:8948–8953

Shastri V, Pishko M (1998) Biomedical applications of electroactive polymers. In: Wise D, Wnek G, Trantolo D, Gresser J (eds) Electrical and optical systems: fundamentals, methods and applications. World Scientific Publishing Company, New York, pp 1031–1051

Shastri V, Schmidt C, Kim T-H, Vacanti J, Langer R (1996) Polypyrrole-A potential candidate for stimulated nerve regeneration. Materials Research Society Meeting 414:113–117

Shastri V, Rahman N, Martin I, Langer R (1999) Applications of conductive polymers in bone regeneration. Mat Res Soc Symp Proc 550:215–219

Shastri V, Martin I, Langer R (2000) Macroporous polymer foams by hydrocarbon templating. Proc Natl Acad Sci USA 97(5):1970–1975

Shi Q, Rafii S, Wu MH, Wijelath ES, Yu C, Ishida A, Fujita Y, Kothari S, Mohle R, Sauvage LR, Moore MA, Storb RF, Hammond WP (1998) Evidence for circulating bone marrow-derived endothelial cells. Blood 92(2):362–367

Sittinger M, Bujia J, Minuth W, Hammer C, Burmester G (1994) Engineering of cartilage tissue using bioresorbable polymer carriers in perfusion culture. Biomaterials 15(6):451–456

Springer M, Chen A, Kraft P, Bednarski M, Blau H (1998) VEGF gene delivery to muscle: potential role for vasculogenesis in adults. Mol Cell 2:549–558

Steinberg MS (1963) Reconstruction of tissues by dissociated cells. Science 141:401–408

Taylor D, Zane Atkins B, Hungspreugs P, Jone T, Reedy M, Hutcheson K, Glower D, Kraus W (1998) Regenerating functional myocardium: improved performance after skeletal myoblast transplantation. Nat Med 4(8):929–933

Thomson JA, Itskovitz-Eldor J, Shapiro SS, Waknitz MA, Swiergiel JJ, Marshall VS, Jones JM (1998) Embryonic stem cell lines derived from human blastocysts. Science 282:1145–1147

Vacanti C, Langer R, Schloo B, Vacanti J (1991) Synthetic polymers seeded with chondrocytes provide a template for new cartilage formation. Plast Reconstr Surg 88(5):753–759

Vacanti J, Langer R (1999) Tissue engineering: the design and fabrication of living replacement devices for surgical reconstruction and transplantation. Lancet 354 [Suppl 1]:SI32–4

Vacanti J, Morse M, Saltzman W, Domb A, Perez-Atayde A, Freed L, Langer R (1988) Selective cell transplantation using bioabsorbable artificial polymers as matrices. J Pediatr Surg 23(1 Pt):23–29

van Wachem, P, Stronck J, Koers-Zuideveld R, Dijk F, Wildevuur C (1990) Vacuum cell seeding: a new method for the fast application of an evenly distributed cell layer on porous vascular grafts. Biomaterials 11(8):602–606

Villa A, Snyder E, Vescovi A, Martinez-Serrano A (2000) Establishment and properties of a growth factor-dependent, perpetual neural stem cell line from the human CNS. Exp Neurol 161(1):67–84

Vunjak-Novakovic, G, Obradovic B, Martin I, Bursac P, Langer R, Freed L (1998) Dynamic cell seeding of polymer scaffolds for cartilage tissue engineering. Biotechnol Prog 14(2):193–202

Wald H, Sarakinos G, Lyman M, Mikos A, Vacanti J, Langer R (1993) Cell seeding in porous transplantation devices. Biomaterials 14(4):270–278

Wilson J, Jefferson D, Chowdhury J, Novikoff P, Johnston D, Mulligan R (1988) Retrovirus-mediated transduction of adult hepatocytes. Proc Natl Acad Sci USA 85(9):3014–3018

Wilson J, Birinyi L, Salomon R, Libby P, Callow A, Mulligan R (1989) Genetically modified endothelial cells in the treatment of human diseases. Trans Assoc Am Physicians 102:139–147

Yannas I, Burke J (1980) Design of an artificial skin. I. Basic design principles. J Biomed Mater Res 14(1):65–81

Yannas I, Burke J, Gordon P, Huang C, Rubenstein R (1980) Design of an artificial skin. II. Control of chemical composition. J Biomed Mater Res 14(2):107–132

Yannas I, Burke J, Orgill D, Skrabut E (1982) Wound tissue can utilize a polymeric template to synthesize a functional extension of skin. Science 215:174–176

4 Myocardial Tissue Engineering

S. Selbert, W.-M. Franz

1 Heart Transplantation in Germany

Cardiovascular disease and particularly ischemic disorders of the heart are the leading causes of death in Western countries. In 1996 the total mortality in Germany was 882,843; 9.7% died of acute myocardial infarction and 10.7% of congestive heart failure, which is mainly related to ischemic cardiomyopathy occurring after myocardial infarction. Depending on the severity of the disease (New York Heart Association classification) current treatment strategies include bypass operation and percutaneous transluminal coronary angioplasty (PTCA) and stunt implantation. Progressive heart failure related to ischemic cardiomyopathies can ultimately only be treated by transplantation. During the last 5–10 years the rate of heart transplantation in Germany ranged between 478 and 562 transplantations per year (see Internet page http://www.eurotransplant.nl). In consequence, the average waiting time for the patients is between 9 and 12 months, and ultimately 20%–30%

of those patients die before receiving the donor heart. Due to the limited availability of organs, alternative approaches are necessary.

2 Alternatives for Heart Transplantation

2.1 Xenotransplantation

One of the most controversially discussed alternatives for heart transplantation is xenotransplantation. Besides ethical considerations concerning the transplantation of organs between species, currently one of the major barriers to clinically successful xenotransplantation of pig vascularized organs to humans is the antibody- and complement-dependent hyperacute rejection (HAR). Preformed antibodies bind to carbohydrate epitopes on the vascular surface of the xenotransplant (Galα(1,3)Gal). Within seconds and minutes post-transplantation complement activation and subsequent lysis of endothelial cells by pore-forming membrane-spanning complexes lead to thrombosis and the

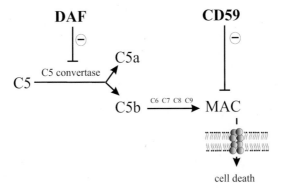

Fig. 1. Transgenic approaches designed to overcome the hyperacute transplant reaction (HAR) of xenotransplants. Transgenic expression of the decay-accelerating factor (DAF) inhibits the C5 convertase catalyzed cleavage of the complement protein C5. Without the generation of the active complement subunit C5b, the membrane-spanning membrane attack complex (MAC) will not be formed and the concomitant cell death can be avoided. A similar approach is the transgenic expression of the glycoprotein CD59, which inhibits the aggregation of MAC and thereby confers protection against complement mediated cell lysis

destruction of foreign tissue (Sandrin and McKenzie 1999). Strategies aimed at preventing HAR involve the inhibition of complement activation by the expression of human cell surface complement regulators such as the decay-accelerating factor (DAF), which inhibits the activation of the C5 convertase, or the glycolipid protein CD59, which at the end of the complement cascade prevents the formation of the cytolytic membrane attack complex (MAC) (Fig. 1).

Transgenic pig hearts expressing, e.g., DAF, were transplanted in primates and an average survival of approximately 40 days was obtained. Further hurdles which also have to be taken are the suppression of acute and chronic cellular responses to pig xenografts and the many, in part yet unforeseeable, problems which are associated with safety aspects regarding xenotransplantation. Lethal viruses like the porcine endogenous retrovirus (PERV) integrate into the pig genomic DNA and may become activated after xenotransplantation and cause AIDS-like syndromes in human (Patience et al. 1997). So far, retrospective surveys have not revealed evidence of PERV infection in any of the 200 patients who have been exposed to living porcine tissue, e.g., for the treatment of burns, extracorporeal blood treatments, and transplantation of islet and neuron cells (Paradis et al. 1999). However, cross-species infection has recently been demonstrated in guinea pigs.

Until these aspects are fully understood, human heart xenotransplantation is unlikely to be performed.

2.2 Cell Transplantation

An alternative approach which is gaining widespread interest is the isolation and implantation of different cell types. Postnatally mature cardiomyocytes are terminally differentiated and do not reenter the cell cycle to a significant extent after injury (Anversa et al. 1991; Soonpaa and Field 1998). Hence, the loss of myocardium after myocardial infarction is irreversible. Since the human heart lacks functional repair mechanisms, skeletal myoblasts, cardiomyocytes, and bone marrow stromal cells have been explored as possible replacement tissues. In animal models it was demonstrated that injected fetal and neonatal cardiomyocytes can stabilize the infarcted myocardium (Li et al. 1996) and revascularization as well as recovery from defective left ventricular function has been observed (Li et al. 1997). On a cellular level, the viable grafts showed N-cadherin-positive adherens junctions and connexin43-positive gap junctions. In some hearts, grafted cardiomyocytes even formed adherens and gap junctions with host cardiomyocytes, suggesting electromechanical coupling between graft and host tissue. More commonly, however, grafts were separated from host myocardium by scar tissue (Reinecke et al. 1999). In addition, it has been shown that transplanted cardiomyocytes can function as carriers for recombinant therapeutic proteins (Koh et al. 1996). As an alternative to cardiomyocytes, skeletal muscle cells have been investigated, and it has been shown that skeletal myoblasts will form differentiated skeletal muscle when transplanted in normal and injured hearts (Taylor et al. 1998; Klug et al. 1996; Leor et al. 1996). A major drawback to the usage of adult skeletal muscle for cardiac repair is its differing electrical properties. Skeletal muscle fibers in contrast to cardiomyocytes are electrically isolated from one another. However, in recent cell-culture experiments it was shown that electromechanical coupling between skeletal and cardiac muscle is possible. When Reinecke et al. (2000) cocultivated neonatal or adult cardiomyocytes with skeletal muscle, they observed that approximately 10% of the skeletal muscle fibers started to contract in synchrony with adjacent cardiomyocytes. Confocal microscopy revealed cadherin and connexin43 at junctions between myotubes and cardiomyocytes. Following dye microinjection, the myotubes transferred the dye to adjacent cardiomyocytes via gap junctions. What remains to be determined is the underlying mechanism and the obstacle to similar junctions being induced in vivo. Recent reports have revealed that bone marrow stromal cells, which have many characteristics of mesenchymal stem cells, may serve as a source for autologous cardiomyocytes (Makino et al. 1999). Following 5-aza-cytidine-induced hypomethylation, bone marrow cells were transplanted into rat ventricular scar tissue. Five weeks after transplantation the researchers were able to detect an induction of angiogenesis, a decrease in transmural scar tissue, and an improvement in peak systolic blood pressure (Tomita et al. 1999).

Efforts to generate myocardial cell lines on the targeted expression of oncogenes, either in cultured cardiomyocytes (Sen et al. 1988) or in the myocardium of transgenic animals (Delcarpio et al. 1991; Katz et al. 1992) have failed. Hence, no established lines exhibiting the typical morphological and functional features of adult cardiomyocytes, in particular after repeated passage, are available. The establishment of human cardiac-specific cell lines, i.e., of ventricular-like cardiomyocytes, would be a major step towards a cell-mediated replacement therapy.

3 Sources for Human Cardiomyocytes

The invention of the nuclear transfer cloning technology demonstrated by the cloning of the sheep Dolly from an udder cell's nucleus (Wilmut et al. 1997) and the derivation of human embryonic stem (ES) cells from human blastocysts (Thomson et al. 1998) have been mentioned regularly as the breakthroughs of the year 1999 (Vogel 1999). Through these inventions it has become feasible to derive stem cells from a patient's own tissue and use those potentially for "tissue regeneration and repair." The pathway to achieve this goal is currently called "therapeutic cloning" and can be described as follows: A cell biopsy is taken from the patient and the nucleus of the somatic cell is transferred into an enucleated donor oocyte with the nuclear transfer techniques pioneered in mice, sheep, and cattle (Wolf et al. 1998). The somatic cell's nucleus, which contains the whole genome, gets reprogrammed and the resulting embryo is allowed to develop until the blastocyst stage. The inner cell mass (ICM) of the blastocyst (see Fig. 2A) is then removed by surgery and cultured, and the ES cells are harvested from it. The ES cells are then directed into the particular cell type required, for example cardiomyocytes, and transplanted into the patient's myocardium. Cardiomyocytes derived by in vitro differentiation of such derived ES cells would be ideal candidates for autologous cell transplantation into the heart. This is the most straightforward approach, but it involves taking a human blastocyst whose capacity for developing into a human being is currently uncertain.

ES cells can be derived from the blastocyst or alternatively from primordial germ cells (PGC) (Fig. 2B), which are cells of the early embryo that eventually differentiate into sperm and oocyte. PGCs themselves can be isolated from the gonadal ridges of aborted fetal material (Shamblott et al. 1998).

Like ES cells, they regenerate seemingly forever, they are pluripotent and can be differentiated in vitro. However, there are strong doubts that

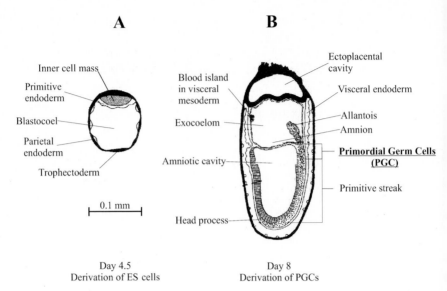

A

Inner cell mass
Primitive endoderm
Blastocoel
Parietal endoderm
Trophectoderm

0.1 mm

Day 4.5
Derivation of ES cells

B

Ectoplacental cavity
Blood island in visceral mesoderm
Visceral endoderm
Exocoelom
Allantois
Amnion
Amniotic cavity
Primordial Germ Cells (PGC)
Primitive streak
Head process

Day 8
Derivation of PGCs

Fig. 2A,B. Sources for the generation of mammalian ES and EG cells. **A** The blastocyst of a mouse embryo at day 4.5 post-conception (p.c.). ES cells are derived from the inner cell mass (ICM) of the preimplantation embryo, in mice at about day 5 and in humans at day 6 after insemination. **B** EG cells are derived from the base of the allantois of an 8-day-old mouse embryo. Human EG cells can be generated between 5 and 9 weeks p.c. They can be distinguished by their large, rounded shape and high levels of alkaline phosphatase activity

EG cells can simply be substituted for ES cells. Recent work presented by Kato et al. (1999) showed that implantation of mouse EG cells into early mouse embryos revealed fetal abnormalities, which contrast with ES cells. This group could further prove that the abnormal development was caused by a lack of DNA methylation, which is called genomic imprinting. Genomic imprints are present in ES cells, egg, sperm, and adult somatic cells, but are temporarily erased in PGCs. It is currently unknown whether the usage of human EG cells may cause developmental abnormalities. Moreover, it is uncertain whether or not genomic imprinting does influence the in vitro differentiation to specific tissues (see Sect. 4 below).

Last but not least it has been shown for the mouse model that mesenchymal bone marrow stromal cells can be forced to generate spontaneously contracting cardiomyocytes (Makino et al. 1999). Already a decade ago similar transdifferentiation experiments in primary fibroblasts and even in ectodermal cells led to the identification and isolation of the muscle determination gene *MyoD* (Choi et al. 1990). However, until now the search for an equivalent "master switch" gene of cardiac determination has not been successful. Transdifferentiation experiments with bone marrow stromal cells frequently give rise to adipocytes, skeletal muscle, and chondrocytes (Dexter et al. 1976). Differentiation of cardiomyocytes from bone marrow stromal cells is emerging and far from any clinical impact. Cardiomyocytes obtained by this pathway are very small in number and so far only available from rodent origin. In contrast to ethical problems connected to human ES and EG cells, this approach might offer a very attractive possibility of obtaining autologous cell transplants from easily accessible bone marrow biopsies.

4 Embryonic Stem Cells

ES cells are derived from the inner cell mass of preimplantation embryos (Doetschman et al. 1985). They can proliferate indefinitely in an undifferentiated state retaining a stable and normal karyotype, are capable of differentiating in vitro and in vivo, and can contribute to the formation of normal tissues and organs of a chimeric individual when injected into a host embryo. It is important to stress that so far only murine ES cells fulfill these criteria. If ES cells are withdrawn from "Leukemia Inhibitory Factor" (LIF) and feeder cells are omitted, differentiation is initiated (Fig. 3). During the differentiation process, ES cells form intermediate structures known as "embryoid bodies" (EBs), which closely resemble the mouse-embryo at day 5.

Differentiating EBs are able to give rise to all of the approximately 210 different cell types, i.e., cardiomyocytes (Doetschman et al. 1985; Wobus et al. 1991) skeletal (Rohwedel et al. 1994) and smooth muscle cells (Drab et al. 1997), neurons (Strubing et al. 1995), and endothelial cells (Wartenberg et al. 1998). Their differentiation pathway can be modulated in vitro by modifying the culture conditions. Our and other

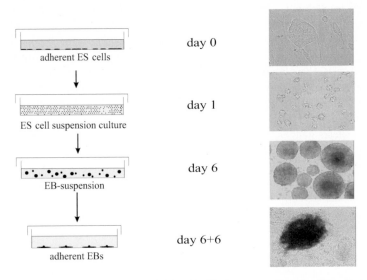

adherent ES cells

day 0

ES cell suspension culture

day 1

EB-suspension

day 6

adherent EBs

day 6+6

Fig. 3. In vitro differentiation of ES cells. Pluripotency of murine ES cells is maintained during adherent cultivation in the presence of the leukemia inhibitory factor (LIF) and fibroblast feeder cells (day 0). Differentiation is induced by non-adherent growth in bacterial plastic dishes omitting LIF and feeder cells. During a time period of 6 days so-called embryoid bodies (EBs) form which resemble the preimplantation embryo. If EBs are allowed to adhere to tissue culture plastic coated with gelatine at day 6+0, they attach and spread out to differentiate all approximately 210 possible cell types (day 6+6)

groups have recently shown that in the presence of micromolar concentrations of retinoic acid (RA), given early during ES-cell differentiation, specifically the cardiomyocytic differentiation pathway is enhanced and the cardiomyocyte development is accelerated. We could also show that ES-cell-derived cardiomyocytes contain all different cardiac lineages: sinusnodal-, Purkinje-, pacemaker-, atrial-, and ventricular-like cells (Wobus et al. 1997). The addition of RA allows a preferential differentiation of cardiomyocytes to ventricular- and Purkinje-like cells. In contrast to other primary or transgenic cardiac cell lines, the ES-cell system provides a valid source for the generation of differentiated human cardiomyocytes, in particular after the recent establishment of human ES-cell lines (Shamblott et al. 1998; Thomson et al. 1998) and their in vitro differentiation (Reubinoff et al. 2000). However, until now the availability of human ES-cell-derived ventricular-like cardiomyocytes is limited by the fact that in Germany the generation of human ES cells is still prohibited by law. At present, experimental work can only be done with single characterized ES-cell lines, which are commercially available overseas. The maintenance and differentiation of pre-existing human ES-cell lines is currently a focus of public discussion.

5 Isolation of In Vitro-Differentiated Cardiomyocytes

Undifferentiated ES cells when transplanted subcutaneously give rise to teratocarcinomas. ES-cell differentiation as well as the in vitro differentiation of other pluripotent cells gives rise to a mixture of various cell types, among which about 5%–20% of cardiac-like cells are present. Having clinical trials in mind a pure population of non-oncogenic cardiomyocytes is mandatory. Therefore, the small number of cardiomyocytes in the EB needs to be purified from the majority of non-cardiomyogenic cells. Assuming that the transplantation of pacemaker-like cardiomyocytes into the host myocardium will cause arrhythmia, the challenge becomes even higher, raising the question: How is it possible to purify a specific subtype of a population of cardiomyocytes out of the EB?

The solution for this problem is the specific labeling of cardiomyocytes by a marker gene under the transcriptional control of a cardiac-specific promoter and the concomitant purification (Müller et al. 2000) (Fig. 4).

5.1 Labeling of Cardiomyocytes

In order to label ventricular-like cardiomyocytes, we cloned the enhanced green fluorescent protein (EGFP)-gene under the transcriptional control of the previously characterized ventricular myosin light chain 2 (MLC2v)-promoter (Franz et al. 1993). In transgenic animal models, the ventricular specificity of the cardiac 2.1 kb MLC2v-promoter has been demonstrated (Franz et al. 1993). So far, the MLC2v promoter is the only available ventricle-specific promoter with no extracardiac activity

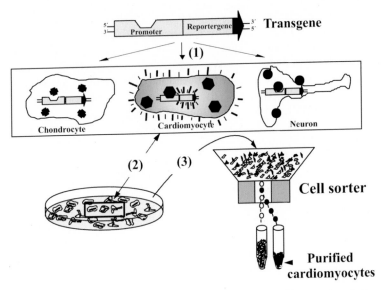

Fig. 4. Labeling and selection of in vitro differentiated cardiomyocytes. A tissue-specific promoter drives the expression of a reporter gene, which allows the labeling of a specific cell type (*1*). Following the in vitro differentiation of ES cells (*2*) fluorescence-labeled cardiomyocytes can be selected by fluorescent-activated cell sorting FACS (*3*)

(see Table 1). Using transgenic approaches, other promoters were characterized for their cardiac activity. These promoters can be grouped as follows:

– Promoters that are active both in atrium and in ventricle, such as α-MHC promoter (Subramaniam et al. 1991)
– Promoters with atrium-specific activity such as the atrial natriuretic factor (ANF) promoter (Seidman et al. 1991)
– Promoters with ventricle-specific activity such as the MLC-2 promoter (Franz et al. 1993)
– Promoters that are active in conductive tissue, such as the troponin I promoter (Zhu et al. 1995)

Table 1. Promoters with cardiac-specific activity in transgenic mice

Tissue specificity	Coexpression	Reference
Atrium		
Atrial natriuretic factor (ANF)	Hypothalamus	Seidman et al. 1991
A1 adenosine receptor (A1AR)	Brain	Rivkees et al. 1999
Ventricle		
Myosin light chain 2 (MLC2v)	None	Franz et al. 1993
β-Myosin heavy chain (β-MHC)	Skeletal muscle	Rindt et al. 1993
Dystrophin	None	Kimura et al. 1997
	(no expression in left ventricle)	
Atrium and ventricle		
α-Myosin heavy chain (α-MHC)	Lung	Subramaniam et al. 1991
α-Skeletal actin	Skeletal muscle, stomach	Brennan and Hardeman 1993
M-Creatine kinase (CK-M)	Skeletal muscle, lung, brain	Johnson et al. 1989
Myoglobin	Skeletal muscle	Parsons et al. 1993
Myosin light chain 3F (MLC-3F)	Skeletal muscle	Kelly et al. 1995
α-B-Crystallin	Skeletal muscle	Gopal-Srivastava et al. 1995
Cardiac actin	Skeletal muscle	Biben et al. 1996
Adenylosuccinate-synthetase	Skeletal muscle, blood vessels	Lewis et al. 1999
Phospholamban	None	McTiernan et al. 1999
Conductive tissue (AV node)		
Troponin I	Skeletal muscle, cartilage	Zhu et al. 1995

In Table 1 only the promoters of the ventricular MLC-2, the phospholamban (PLB), and the dystrophin gene show no ectopic expression in other tissues than the heart. The transgenic experiments are the gold standard to prove heart specificity. It is important to mention that all the listed promoters have their own characteristic expression pattern depending on the regulatory elements present on the truncated DNA. While the PLB promoter is active in the entire heart, the promoter activity of the dystrophin promoter is restricted to the right heart. It should be noted that the promoters summarized in Table 1 differ from each other in their overall strength, and that they show different activity profiles during embryonal development (Franz et al. 1993). In recent experiments aimed at the purification of in vitro-differentiated cardiomyocytes, the tissue-specific alpha myosin heavy chain (αMHC)-promoter was used to drive the antibiotic resistance gene aminoglycoside phosphotransferase (Neo^R) thereby conferring neomycin resistance selectively to in vitro-differentiating cardiomyocytes (Klug et al. 1996). Even so, this experiment yielded pure cardiomyocyte-like cells; after transplantation these cells caused tumor formation. In addition, it might be speculated that taking the in vitro-developing cardiomyocytes out of their natural environment of the EB for an extended period of time, as happens during the antibiotic selection process, dedifferentiation processes would be triggered. This phenomenon can be regularly seen during the cultivation of primary cardiomyocytes. We used the green fluorescent protein (EGFP)-marker gene under the transcriptional control of the MLC2v-promoter to specifically label in vitro-generated cardiomyocytes (Müller et al. 2000). In contrast to a selection protocol using antibiotic selection, our procedure has the advantages that the cardiomyocytes are kept in their native environment for as long as possible and the labeled cells can be visualized and thereby be characterized in vivo. In addition, the fluorescent cells can be purified in a fast protocol at any given time. We transfected the MLC2v-EGFP construct in undifferentiated ES cells and isolated ES-cell clones which contained the cardiomyocyte-specific expression vector. Among those, we identified clones which after 14 days of differentiation showed first green fluorescent cardiomyocytes which were spontaneously contracting. In the following days the number of EGFP-positive cells further increased, and at day 30 a maximum of about 5%–10% of fluorescent cardiomyocytes were visible (Fig. 5).

Our expectation that the green fluorescent cells were of a ventricular-like phenotype were corroborated by immunohistochemical, electrophysiological, and pharmacological measurements. As expected, EGFP-labeled cells contained α-actinin-positive sarcomeres and were negative for skeletal muscle markers like fast skeletal myosin. In patch clamp analysis, 82% of EGFP-positive cells displayed typical ventricular-like action potentials (AP) with a negative membrane potential of -68.6 ± 2.8 mV, an AP duration (APD) of 118.3 ± 15.2 ms, an overshoot of 34.3 ± 3.9 mV and the clearly pronounced plateau phase indicating the long-lasting Ca^{2+} influx through l-type Ca^{2+} channels (Maltsev et al. 1999) (Fig. 6A). Treatment of these cells with the β-adrenergic agonist isoprenaline (Iso, 0.1 μM) led to a prolongation of the APD, which is

Fig. 5. Green fluorescent labeling of in vitro-differentiated ventricular-like cardiomyocytes. The picture shows part of a spontaneously contracting area of an EB at day 6+10. Approximately 25% of the beating cells are green fluorescent

known to prolong APD due to stimulation of l-type Ca^{2+}-channels (Maltsev et al. 1999). Incubation with the muscarinic agonist carbachol (CCh, 1 μM) which has a pronounced negative chronotropic effect on the spontaneous electrical activity, did not result in any effect on APD and the membrane potential of EGFP-positive cells as can be expected for ventricular-like cardiomyocytes. Of EGFP-positive cells, 12% showed atrial-like membrane potentials while 6% could be classified as early type cardiomyocytes. In contrast, EGFP-negative cells revealed a strong negative chronotropic response upon CCh application without concomitant hyperpolarization (Fig. 6B).

In summary ventricular-like cardiomyocytes can be generated in vitro and can be labeled and characterized using MLC2v driven fluorescent marker genes.

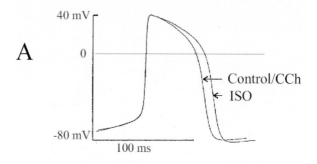

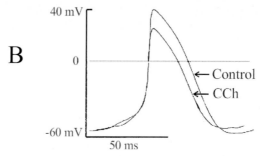

Fig. 6. Electrophysiological analysis of spontaneously contracting cardiomyocytes. Patch clamp recordings revealed that 82% of EGFP-positive cells were of a typical ventricular-like phenotype (**A**), while EGFP-negative cardiomyocytes (**B**) showed characteristics typical for early cardiomyocytes

5.2 Purification of ES-Cell-Derived Cardiomyocytes

To obtain a pure fraction of ventricular cardiomyocytes, the green fluorescent cells are subjected to fluorescent-activated cell sorting (Fig. 4). But before that, a single cell suspension has to be prepared. For that reason, the EBs are gently incubated in a mixture of proteinases which brake down cell–cell contacts and degrade the extracellular matrix without causing any significant cell damage. To enrich the fraction of cardiomyocytes within the crude cell population, the whole cell suspension is laid on a percoll density gradient. During centrifugation,

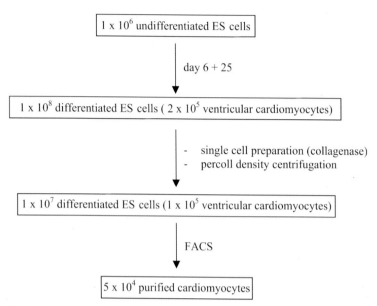

Fig. 7. Enrichment of ventricular-like cardiomyocytes generated from murine ES cells

cardiomyocytes preferentially accumulate at the interphase of the gradient and thereby the concentration of cardiomyocytes increases three- to fivefold. The final FACS sorting separates fluorescent ventricular-like cardiomyocytes from the majority of non-fluorescent cells. The result is a population of ventricular-like cardiomyocytes which is about 97% pure. The isolated cardiomyocytes immunohistochemically show sarcomeric structures, are beating spontaneously, and are ready for transplantation into the diseased heart. The protocol described gives reproducible results and yields 5×10^4 EGFP-positive cardiomyocytes starting up with 1×10^6 undifferentiated or 1×10^8 differentiated ES cells (see Fig. 7).

6 Outlook

Using the above protocol, murine ES-cell-derived cardiomyocytes can be labeled and purified efficiently and in quantities which are sufficient for animal transplantations or drug development. Having patients and human ES cells in mind, several problems have to be solved before an equivalent protocol can be designed for the isolation of human cardiomyocytes.

– Methods for the in vitro differentiation of human ES cells are just emerging, and first experiments in this direction showed that human ES cells in contrast to their murine counterparts seem to be insensitive to LIF, which means it might be difficult to maintain their pluripotent phenotype (Reubinoff et al. 2000).
– Current cardiomyocyte labeling and purification protocols are dependent on non-human transgenes, such as EGFP or the neomycin-resistance gene. Following transplantation into the recipient myocardium, it can be taken for granted that transgene-positive transplants presenting the foreign gene will provoke an immunoreaction which finally leads to the elimination of all transplanted cardiomyocytes. To suppress transplant rejection, either the recipient has to be treated with immunosuppressive agents or all non-human transgenes have to be eliminated before being injected into the human system (e.g., by using inducible sequence-specific Cre recombination technology).

Five years ago nobody would have believed that autologous cell transplantation would ever be possible in the human myocardium. In the meantime, all the tools, such as human ES cells and the nuclear transfer technology, are at hand to make this dream become true. It now seems only a matter of time until the first human in vitro-differentiated cardiomyocytes are at hand and will be transplanted into a recipient heart. After all, the most important issue to successfully continue with this kind of medical research is that all decisions made are in consensus with ethical values.

References

Anversa P, Fitzpatrick D, Argani S, Capasso M (1991) Myocyte mitotic division in the aging mammalian rat heart. Circ Res 69:1159–1164

Biben C, Hadchouel J, Tajbakhsh S, Buckingham M (1996) Developmental and tissue-specific regulation of the murine cardiac actin gene in vivo depends on distinct skeletal and cardiac muscle-specific enhancer elements in addition to the proximal promoter. Dev Biol 173:200–212

Brennan KJ, Hardeman EC (1993) Quantitative analysis of the human alpha-skeletal actin gene in transgenic mice. J Biol Chem 268:719–725

Choi J, Costa ML, Mermelstein CS, Chagas C, Holtzer S, Holtzer H (1990) MyoD converts primary dermal fibroblasts, chondroblasts, smooth muscle, and retinal pigmented epithelial cells into striated mononucleated myoblasts and multinucleated myotubes. Proc Natl Acad Sci USA 87:7988–7992

Delcarpio JB, Lanson NA, Field LJ, Claycomb WC (1991) Morphological characterization of cardiomyocytes isolated from a transplantable cardiac tumor derived from transgenic mouse atria (AT-1 cells). Circ Res 69:1591–1600

Dexter TM, Allen TD, Lajtha LG (1976) Conditions controlling the proliferation of haemopoietic stem cells in vitro. J Cell Physiol 91:335–344

Doetschman TC, Eistetter H, Katz M, Schmidt W, Kemler R (1985) The in vitro development of blastocyst-derived embryonic stem cell lines: formation of visceral yolk sac, blood islands and myocardium. J Embryol Exp Morphol 87:27–45

Drab M, Haller H, Bychkov R, Erdmann B, Lindschau C, Haase H, Morano I, Luft FC, Wobus AM (1997) From totipotent embryonic stem cells to spontaneously contracting smooth muscle cells: a retinoic acid and db-cAMP in vitro differentiation model. Faseb J 11:905–915

Franz WM, Breves D, Klingel K, Brem G, Hofschneider PH., Kandolf R (1993) Heart-specific targeting of firefly luciferase by the myosin light chain-2 promoter and developmental regulation in transgenic mice. Circ Res 73:629–638

Gopal-Srivastava R, Haynes JH, Piatigorsky J (1995) Regulation of the murine αB-crystallin/small heat shock protein gene in cardiac muscle. Mol Cell Biol 15:7081–7090

Johnson JE, Wold BJ, Hauschka SD (1989) Muscle creatine kinase sequence elements regulating skeletal and cardiac muscle expression in transgenic mice. Mol Cell Biol 9:3393–3399

Kato Y, Rideout III WM, Hilton K, Barton SC, Tsunoda Y, Surani MA (1999) Developmental potential of mouse primordial germ cells. Development 126:1823–1832

Katz EB, Steinhelper ME, Delcarpio JB, Daud AI, Claycomb WC, Field LJ (1992) Cardiomyocyte proliferation in mice expressing alpha-cardiac myosin heavy chain-SV40 T-antigen transgenes. Am J Physiol 262:H1867–H1876

Kelly R, Alosno S, Tajbakhsh S, Cossu G, Buckingham M (1995) Myosin light chain 3F regulatory sequences confer regionalized cardiac and skeletal muscle expression in transgenic mice. J Cell Biol 129:383–396

Kimura S, Abe K, Suzuki M, Ogawa M, Yoshioka K, Kaname T, Miike T, Yamamura K (1997) A 900 bp genomic region from the mouse dystrophin promoter directs lacZ reporter expression only to the right heart of transgenic mice. Dev Growth Differ 39:257–265

Klug MG, Soonpaa MH, Koh GY, Field LJ (1996) Genetically selected cardiomyocytes from differentiating embronic stem cells form stable intracardiac grafts. J Clin Invest 98:216–224

Koh GY, Kim SJ, Klug MG, Park K, Soonpaa MH, Field LJ (1996) Targeted expression of transforming growth factor $\beta1$ in intercardiac grafts promotes vascular endothelial DNA synthesis. J Clin Invest 95:114–121

Leor J, Patterson M, Quinones MJ, Kedes LH, Kloner RA (1996) Transplantation of fetal myocardial tissue into the infarcted myocardium of rat. A potential method for repair of infarcted myocardium? Circulation 94:II332–II336

Lewis AL, Xia Y, Datta SK, McMillin J, Kellems RE (1999) Combinatorial interactions regulate cardiac expression of the murine adenylosuccinate synthetase 1 gene. J Biol Chem 274:14188–14197

Li RK, Jia ZQ, Weisel RD, Mickle DA, Zhang J, Mohabeer MK, Rao V, Ivanov J (1996) Cardiomyocyte transplantation improves heart function. Ann Thorac Surg 62(3):654–660

Li RK, Mickle DA, Weisel RD, Mohabeer MK, Zhang J, Rao V, Li G, Merante F, Jia ZQ (1997) Natural history of fetal rat cardiomyocytes transplanted into adult rat myocardial scar tissue. Circulation. 96[9 Suppl]:II-179–186

Makino S, Fukuda K, Miyoshi S, Konishi F, Kodama H, Pan J, Sano M, Takahashi T, Hori S, Abe H, Hata J, Umezawa A, Ogawa S (1999) Cardiomyocytes can be generated from marrow stromal cells *in vitro*. J Clin Invest 103:697–705

Maltsev VA, Ji GJ, Wobus AM, Fleischmann BK, Hescheler J (1999) Establishment of beta-adrenergic modulation of L-type Ca2+ current in the early stages of cardiomyocyte development. Circ Res 84:136–145

McTiernan CF, Lemster BH, Frye CS, Johns DC, Feldman AM (1999) Characterization of proximal transcription regulatory elements in the rat phospholamban promoter. J Mol Cell Cardiol 31:2137–2153

Müller M, Fleischmann BK, Selbert S, Ji GJ, Endl E, Middeler G, Muller OJ, Schlenke P, Frese S, Wobus AM, Hescheler J, Katus HA, Franz WM (2000)

Selection of ventricular-like cardiomyocytes from ES cells in vitro. FASEB J 14(15):2540–2548

Paradis K, Langford G, Long Z, Heneine W, Sandstrom P, Switzer WM, Chapman LE, Lockey C, Onions D, Otto E (1999) Search for cross-species transmission of porcine endogenous retrovirus in patients treated with living pig tissue. Science 285:1236–1241

Parsons WJ, Richardson JA, Graves KH, Williams RS, Moreadith RW (1993) Gradients of transgene expression directed by the human myoglobin promoter in the developing mouse heart. Proc Natl Acad Sci USA 90:1726–1730

Patience C, Takeuchi Y, Weiss A (1997) Infection of human cells by endogenous retrovirus of pigs. Nat Med 3:282–286

Reinecke H, Zhang M, Bartosek T, Murry CE (1999) Survival, integration, and differentiation of cardiomyocyte grafts: a study in normal and injured rat hearts. Circulation 100:193–202

Reinecke H, MacDonald GH, Hauschka SD, Murry CE (2000) Electromechanical coupling between skeletal and cardiac muscle. Implications for infarct repair. J Cell Biol 149(3):731–740

Reubinoff BE, Pera MF, Fong C-Y, Trounson A, Bongso A (2000) Embryonic stem cell lines from human blastocysts: somatic differentiation in vitro. Nat Biotech 18:399–404

Rindt H, Gulick J, Knotts S, Neumann J, Robbins J (1993) In vivo analysis of the murine beta-myosin heavy chain gene promoter. J Biol Chem 268:5332–5338

Rivkees SA, Chen M, Kulkarni J, Browne J, Zhao Z (1999) Characterization of the murine A1 adenosine receptor promoter, potent regulation by GATA-4 and Nkx2.5. J Biol Chem 274:14204–14209

Rohwedel J, Maltsev V, Bober E, Arnold HH, Hescheler J, Wobus AM (1994) Muscle cell differentiation of embryonic stem cells reflects myogenesis in vivo: developmentally regulated expression of myogenic determination genes and functional expression of ionic currents. Dev Biol 164:87–101

Sandrin MS, McKenzie IFC (1999) Recent advances in xenotransplantation. Cur Opin Immunol 11:527–531

Seidman CE, Schmidt EV, Seidman JG (1991) cis-Dominance of rat atrial natriuretic factor gene regulatory sequences in transgenic mice. Can J Physiol Pharmacol 69:1486–1492

Sen A, Dunnmon P, Henderson SA., Gerard RD, Chien KR (1988) Terminally differentiated neonatal rat myocardial cells proliferate and maintain specific differentiated functions following expression of SV40 large T antigen. J.Biol Chem 263:19132–19136

Shamblott MJ, Axelman J, Wang S, Bugg EM, Littlefield JW, Donovan PJ, Blumenthal PD, Huggins GR, Gearhart JD (1998). Derivation of pluripotent stem cells from cultured human primordial germ cells. Proc Natl Acad Sci USA 95:13726–13731

Soonpaa MH, Field LJ (1998) Survey of studies examining mammalian cardiomyocyte DNA synthesis. Circ Res 83:15–26

Strubing C, Ahnert-Hilger G, Shan J, Wiedenmann B, Hescheler J, Wobus AM (1995) Differentiation of pluripotent embryonic stem cells into the neuronal lineage in vitro gives rise to mature inhibitory and excitatory neurons. Mech Dev 53:275–287

Subramaniam A, Jones WK, Gulick J, Wert S, Neumann J, Robbins J (1991) Tissue-specific regulation of the α-myosin heavy chain gene promoter in transgenic mice. J Biol Chem 266:24613–24620

Taylor DA, Atkins BZ, Hungspreugs P, Jones TR, Reedy MC, Hutcheson KA, Glower DD, Kraus WE (1998) Regenerating functional myocardium: Improved performance after skeletal myoblast transplantation. Nat Med 4:929–933

Thomson JA, Itskovitz-Eldor J, Shapiro SS, Waknitz MA, Swiergiel JJ, Marshall VS, Jones JM (1998) Embryonic stem cell lines derived from human blastocysts. Science 282:1145–1147

Tomita S, Li RK, Weisel RD, Mickle DA, Kim EJ, Sakai T, Jia ZQ (1999) Autologous transplantation of bone marrow cells improves damaged heart function. Circulation 100:II247–II256

Vogel G (1999) Breakthrough of the year: capturing the promise of youth. Science 286:2238–2239

Wartenberg M, Gunther J, Hescheler J, Sauer H (1998) The embryoid body as a novel in vitro assay system for antiangiogenic agents. Lab Invest 78:1301–1314

Wilmut I, Schnieke AE, McWhir J, Kind AJ, Campbell KH (1997) Viable offspring derived from fetal and adult mammalian cells. Nature 385:810–813

Wobus AM, Wallukat G, Hescheler J (1991) Pluripotent mouse embryonic stem cells are able to differentiate into cardiomyocytes expressing chronotropic responses to adrenergic and cholinergic agents and Ca2+ channel blockers. Differentiation 48:173–182

Wobus AM, Kaomei G, Shan J, Wellner MC, Rohwedel J, Ji G, Fleischmann B, Katus HA, Hescheler J, Franz WM (1997) Retinoic acid accelerates embryonic stem cell-derived cardiac differentiation and enhances development of ventricular cardiomyocytes. J Mol Cell Cardiol 29:1525–1539

Wolf E, Zakhartchenko V, Brem G (1998) Nuclear transfer in mammals: recent developments and future perspectives. J Biotechnol 65:99–110

Zhu L, Lyons GE, Juhasz O, Joya JE, Hardeman EC, Wade R (1995) Developmental regulation of troponin I isoform genes in striated muscles of transgenic mice. Dev Biol 169:487–503

5 Cellular Transplantation for the Treatment of Heart Failure

P. Menasché

1 Introduction

Because they are terminally differentiated cells, adult cardiomyocytes cannot regenerate and there is no myocardial pool of stem cells to replace those which have suffered irreversible ischaemic injury. The responses to such an injury involve evolution of the infarct zone toward a fibrous noncontractile scar and hypertrophy of cells harboured in the still-viable segments of the heart. At best, these compensatory responses can temporarily maintain an adequate contractile function. At worst, the combination of interstitial fibrosis and inappropriate remodelling promote deterioration of systolic and diastolic functions and lead to heart failure.

This condition has now become a major problem of public health because of its prevalence (approximately 5 million patients in the U.S.), incidence (300,000–500,000 cases per year), high mortality (up to 30%) and the related costs due to drugs and repeated hospitalizations. When heart failure becomes refractory to medical therapy, patients can be offered a wide variety of surgical therapies which culminate in cardiac transplantation. However, organ shortage still results in a substantial mortality (approximately 30%) while on the waiting list; on the other hand, dynamic cardiomyoplasty and partial resections of the left ventricle have yielded overall disappointing clinical results, whereas permanently implanted assist devices are still in a developmental stage. Thus, there is clearly room left for alternate therapies and, over the past decade, experience suggests that transplantation of contractile cells into the myocardium might be one of them.

2 Fetal Cardiomyocytes

Initial experiments with fetal cardiomyocytes, conducted by Soonpa and co-workers (1994) in transgenic mice expressing the gene of β-galactosidase have shown that fetal cardiomyocytes could form stable intramyocardial grafts for up to 2 months and develop connexins, identified as intercalated discs, with host cardiomyocytes. Additional studies conducted in animals suffering from Duchenne's muscular dystrophy (which has allowed tracing of the dystrophin-positive injected cells) have supported these data by demonstrating gap junctions between transplanted fetal cardiomyocytes and the surrounding host myocardium.

An important step has then been the demonstration that these morphological patterns translated into improved functional outcomes. Thus, Li and co-workers (1996) have transplanted fetal cardiomyocytes into fibrotic scars created by cryoinjury in rats and demonstrated that, two months later, the Langendorff-perfused hearts of these animals developed better functional indices than controls only receiving the culture medium. Likewise, in our group, Scorsin and co-workers (1997) have used a rat model of reperfused infarct and shown that injection of fetal cardiomyocytes 30 min after the onset of reperfusion resulted in an increase in ejection fraction and cardiac output, as assessed by echocardiography 1 month later. Of note, a recent study by Li and his associates (2000) comparing three types of fetal cells (cardiomyocytes, smooth muscle cells and fibroblasts) has clearly established the functional superiority of the cardiomyocyte cell line, thereby supporting the concept that the intrinsic contractile properties of the grafted cells are critical for the procedure to be functionally optimized.

However, in a clinical perspective, the use of fetal cells raises several issues associated with ethics, rejection and procurement, which has motivated the interest for another type of contractile cells, i.e. satellite cells or skeletal myoblasts.

3 Skeletal Myoblasts

Satellite cells are normally present in a quiescent state under the basal membrane of muscular fibres. After an injury, they rapidly mobilize, actively proliferate and ultimately fuse to regenerate the damaged myotubes. Several of their characteristics make them attractive for use in the setting of clinical cellular transplantation: (1) an autologous origin, which overcomes problems associated with availability and immunology, (2) the ease with which they can be expanded in vitro so that the problem of scale-up can be satisfactorily addressed, (3) their commitment to an exclusively myogenic differentiation, which should guarantee against tumorigenicity, (4) a high resistance to ischaemia, reflected by their survival after engraftment in nonperfused fibrotic scars, and (5) a potential for self-regeneration, which raises the attractive albeit still purely speculative hypothesis that if the new intramyocardial myotubes happen to be damaged by a recurrent ischaemic event, some regeneration might still be possible from this pool of precursors.

The functional benefits of autologous skeletal myoblast transplantation have first been demonstrated by Taylor and co-workers (1998) in a rabbit model of cryonecrosis. Left ventricular function, assessed 6 weeks after the procedure by micromanometry and sonomicrometry, was found to have significantly improved in the transplanted group compared with control rabbits and those in which grafted cells could not be detected, thereby strongly suggesting the causal relationship between the presence of myoblasts and the functional benefits. Subsequent studies by the same group (Atkins et al. 1999) have then pointed out that the post-transplant improvement primarily involved diastolic function, whereas that of systolic performance was less consistent. In our group (Scorsin et al. 2000), we have compared skeletal myoblasts and fetal cells and found that, after 1 month, the echocardiographically assessed left ventricular ejection fraction was improved to a similar extent in the two groups, with myoblasts being identified in all grafted animals by positive staining for myosin heavy chain. A subsequent study (Pouzet et al. 2001) then established that the functional improvement yielded by skeletal myoblast transplantation was linearly related to the number of injected cells but was independent of the baseline ejection fraction, i.e. was still present in case of severely depressed post-infarct left ventricular function (ejection fraction .25), thereby suggesting that myoblast transplantation could be relevant to the most severe forms of ischaemic heart failure. The clinical relevance of the procedure is further strengthened by our more recent observations (unpublished data) that the benefits of autologous myoblast transplantation are additive to those of angiotensin-converting enzyme inhibitors. Importantly, all these findings made in rats and rabbits have now been confirmed in large animal models involving dogs (Rao et al. 2000) and sheep (data on file), and no study has yet reported that the foci of engrafted cells were arrhythmogenic.

Although there is consistent evidence that skeletal myoblast transplantation improves post-infarct left ventricular function, several basic issues remain unanswered. First, it is unclear whether, and how, grafted myoblasts interact with host cardiomyocytes. Histological studies by Chiu et al. (1995) and Taylor et al. (1998) have shown, in the core of the transplanted area (in the dog and rabbit hearts, respectively) structures resembling cardiac-specific intercalated discs. Conversely, we have failed to detect a positive staining for connexin-43, which is the major gap junction protein. This observation is consistent with those of Reinecke et al. (2000) who have reported that only undifferentiated rat skeletal myoblasts express N-cadherin (the major adhesion protein of the intercalated disc) and connexin-43, whereas these proteins are markedly down-regulated after differentiation into myotubes, both in culture and after myoblast implantation in normal or cryoinjured hearts. One possibility is that coupling develops through connexins different from connexin-43; alternatively, one can hypothesize that even if the grafted myoblasts remain electrically insulated, they might respond to the mechanical stimulation exerted by the surrounding myocardium and thus beat synchronously with host cardiomyocytes. Clearly, this area requires further investigations to better understand the mechanism by which myoblast transplantation improves function. A second issue pertains to the possible phenotypic changes of the grafted myoblasts over time. Chiu et al. (1995) have advocated the theory of a "milieu-induced" differentiation, which would cause myoblasts to acquire a cardiac-like phenotype. The observations of Atkins et al. (1999) partly support this concept in that these authors have observed, at the periphery of cryoinfarcts, clusters of non-skeletal (myogenin-negative) muscle cells resembling immature cardiomyocytes, but these phenotypic patterns still require confirmation. The potential impact of transplanted cells on the induction of angiogenesis is another issue which remains unaddressed. Elucidation of these various problems should help in defining the mechanism(s) responsible for the functional benefits of skeletal myoblast transplantation, and which possibly include direct contribution to contractility, limitation of ventricular remodelling due to the elastic properties of the grafted cells and release of growth and/or angiogenic factors.

Notwithstanding these problems related to basic research, it appears that the functional benefits of skeletal myoblast transplantation have been demonstrated convincingly enough to consider phase I clinical trials. This, in turn, raises specific problems related to both regulatory and technical constraints. The former require cell cultures to be performed in GMP facilities, with clinical-grade media and additives. Technically, scale-up techniques have to be developed to yield large number of cells, among which a high percentage of myoblasts, within the shortest time period so that cell transplantation can be performed reasonably close to the preceding muscular biopsy. Our recent efforts to make this transition "from bench to bedside" should allow us to start using autologous skeletal myoblast transplantation clinically in the near future with safety and feasibility as primary end points. At a later stage, proper identification of the most suitable candidates for this procedure, definition of efficacy end points and assessment of the optimal approach for cell delivery (intraoperative vs endoventricular) will have to be worked out. In addition, in parallel to the early clinical trials, it is worth investigating other cell types, in particular bone marrow stromal cells as preliminary studies (Makino et al. 1999) suggest that they might, under certain culture conditions, differentiate into contractile cells while they share with skeletal myoblasts a potential use as autografts.

72 P. Menasché

References

Atkins BZ, Hueman MT, Meuchel JM, Cottman MJ, Hutcheson KA, Taylor DA (1999a) Myogenic cell transplantation improves in vivo regional performance in infarcted rabbit myocardium. J Heart Lung Transplant 18:1173–1180

Atkins BZ, Lewis CW, Kraus WE, Hutcheson KA, Glower DD, Taylor DA (1999b) Intracardiac transplantation of skeletal myoblasts yields two populations of striated cells in situ. Ann Thorac Surg 67:124–129

Chiu RC-J, Zibaitis A, Kao RL (1995) Cellular cardiomyoplasty: myocardial regeneration with satellite cell implantation. Ann Thorac Surg 60:12–18

Li R-K, Jia Z-Q, Weisel RD, Mickle DAG, Zhang J, Mohabeer MK, Rao V, Ivanov J (1996) Cardiomyocyte transplantation improves heart function. Ann Thorac Surg 62:654–661

Makino S, Fukuda K, Miyoshi S, Konishi F, Kodama H, Pan J, Sano M, Takahashi T, Hori S, Abe H, Hata J, Umezawa A, Ogawa S (1999) Cardiomyocytes can be generated from marrow stromal cells in vitro. J Clin Invest 103:697–705

Pouzet B, Vilquin J-T, Messas E, Scorsin M, Fiszman M, Hagège AA, Schwartz K, Menasché P (2001) Factors affecting functional outcome following myoblast cell transplantation. Ann Thorac Surg 71:844–851

Rao RL, Chin TK, Ganote CE, Hossler FE, Li C, Browder W (2000) Satellite cell transplantation to repair injured myocardium. CVR 1:31–42

Reinecke H, McDonald GH, Hauschka SD, Murry CE (2000) Electromechanical coupling between skeletal and cardiuac muscle. Implications for infarct repair. J Cell Biol 149:731–740

Sakai T, Li RK, Weisel RD, Mickle DAG, Jia ZQ, Tomita S, Kim EJ, Yau TM (1999) Fetal cell transplantation: a comparison of three cell types. J Thorac Cardiovasc Surg 118:715–725

Scorsin M, Hagège AA, Marotte F, Mirochnik N, Copin H, Barnoux M, Sabri A, Samuel J-L, Rappaport L, Menasché P (1997) Does transplantation of cardiomyocytes improve function of infarcted myocardium. Circulation 96 [Suppl II]:II188–II193

Scorsin M, Hagège AA, Vilquin J-T, Fiszman M, Marotte F, Samuel J-L, Rappaport L, Schwartz K, Menasché P (2000) Comparison of the effects of fetal cardiomyocytes and skeletal myoblast transplantation on postinfarct left ventricular function. J Thorac Cardiovasc Surg 119:1169–1175

Soonpaa MH, Koh GY, Klug MG, Field LJ (1994) Formation of nascent intercalated disks between grafted fetal cardiomyocytes and host myocardium. Science 264:98–101

Taylor DA, Atkins BZ, Hungspreugs P, Jones TR, Reedy MC, Hutcheson KA, Glower DD, Kraus WE (1998) Regenerating functional myocardium: improved performance after skeletal myoblast transplantation. Nat Med 4:929–933

6 A Truly New Approach for Tissue Engineering: The LOEX Self-Assembly Technique

F.A. Auger, M. Rémy-Zolghadri, G. Grenier, L. Germain

1 Introduction

Tissue engineering has created several original and new avenues in the biomedical sciences. There is ongoing progress, but the tissue-engineering field is currently at a crossroads in its evolution; the validity of this technique is well established. Thus, new clinical applications must appear rapidly, within a few years, so that it will have a true impact on patient care. The self-assembly approach of the Laboratoire d'Organogénèse Expérimentale (LOEX) should be at the forefront.

2 General Principles of the Self-Assembly Approach

One must not forget that tissue engineering was first introduced as a life-saving procedure for burn patients (O'Connor et al. 1981). The successful engraftment of autologous epidermal sheets was the first proof of concept of the powerful technology that we know today (Gallico et al. 1984; Auger 1988; Damour et al. 1997).

The subsequent efforts in the field followed essentially three schools of thought. The first approach consists in the seeding of cells into various gels, which are then reorganized by the incorporated cells (Bell et al. 1979; Weinberg and Bell 1986; L'Heureux et al. 1993; Auger et al. 1995, 1998; Goulet et al. 1997a,b; Germain et al. 1999). Alternatively, a second approach is to seed cells into a scaffold where they will thrive and secrete an extracellular matrix (Berthod et al. 1993; Black et al. 1998; Duplan-Perrat et al. 2000). The scaffold materials are usually bioresorbable over a wide range of time periods depending on their chemical natures (Boyce et al. 1988, 1990; Cooper et al. 1991; Shinoka et al. 1997; Peter et al. 1998; Isogai et al. 1999; Lin et al. 1999; Niklason et al. 1999). A third approach is different since it uses the principle of a tissue template that allows, after implantation, the ingress of cells into the appropriately organized scaffold. Thus, these grafts are acellular and must stimulate the regenerative potential of the tissue in vivo wherever they are implanted (Yannas et al. 1982; Jaksic and Burke 1987; Heimbach et al. 1988; Badylak et al. 1989, 1998; Huynh et al. 1999).

Our group has developed a different and original method for the reconstruction of soft tissues. It takes full advantage of the various intrinsic properties of cells when adequately cultured. This entails some particular media composition and adapted mechanical straining of these three-dimensional structures.

Our own experience with the culture of autologous epidermal sheets gave us some insight into the properties of cells to recreate in vitro a tissue-like structure. Furthermore, our various adaptations of the gel-construct approach has shown that cells could be well aligned, along with their extracellular matrix, if the mechanical forces generated were well applied: either passively by anchorage (López-Valle et al. 1992), or actively by cyclic traction of these constructs (Goulet et al. 1997b).

The self-assembly approach is a combination of all these results. Briefly, we coax cells into secreting their own extracellular matrix in a sheet form. Thereafter, we either roll or stack these sheets to create a three-dimensional human living organ substitute. Our concept has been applied with excellent results to the reconstruction of blood vessel and skin (L'Heureux et al. 1998; Michel et al. 1999).

The living substitutes we created have no biomaterial and the extracellular matrix is not exogenous but corresponds to the cell's optimal need. Such living substitutes have a distinct advantage over other methods because of their superior characteristics in the human body. Such substitutes are then as close as possible to the native tissue.

3 Small-Diameter Tissue-Engineered Blood Vessels

The grafting of commercially available vascular synthetic prostheses is appropriate for the replacement of large arteries (Brewster and Rutherford 1995). However, when long-term patency results of small-diameter blood vessels (<5 mm) are analyzed (limb salvage or coronary bypass surgery for example), they often lead to dismal final results. Indeed, the thrombogenic properties of the biomaterials associated with the low blood flow in these arteries are responsible for the formation of blood clots and obstruction of the conduit in many cases (Stephen et al. 1977; O'Donnell et al. 1984; Sayers et al. 1998). The alternative techniques utilizing allogenic or xenogenically treated implants have given few encouraging results because of the high percentage of prosthesis degradation over time with long-term implantation (Charara et al. 1989; Dardik 1989; Allaire et al. 1994; Stanke et al. 1998).

The first attempt to produce a reconstructed blood vessel by tissue-engineering methods appeared in 1986 with the model presented by Weinberg and Bell (1986). The method put forward by these researchers was based on collagen gels seeded with bovine vascular cells. Such a technique was the basis of subsequent research conducted by other teams (L'Heureux et al. 1993; Ziegler et al. 1995; Tranquillo et al. 1996). But the resulting structures were not resistant enough to sustain normal blood pressure (Weinberg and Bell 1986; L'Heureux et al. 1993) and some of these prostheses had to be reinforced with a synthetic mesh (Weinberg and Bell 1986; Hirai and Matsuda 1996) making them hybrid artificial substitutes (composed of living cells in association with a synthetic support) with all the untoward properties associated with such constructions. Moreover, many of these experiments involved animal cells and sometimes only two layers of blood vessel wall (Hirai and Matsuda 1996; Niklason et al. 1999).

Since the previous approaches did not seem to be conducive to an acceptable clinical result, we developed a tissue-engineered blood vessel (TEBV) based exclusively on the use of human cells in the absence of any synthetic or exogenous material such as animal collagens (L'Heureux et al. 1998). This TEBV was shown to have supra-physiological blood-pressure resistance and a histological organization comparable to that of a native artery.

3.1 From Cells to the Reconstructed Vessel

The cells used for such a TEBV were endothelial cells (EC) and smooth muscle cells (SMC) isolated from human umbilical cord veins using an enzymatic method for EC (Jaffe et al. 1973) and the method of Ross for SMC isolation (Ross 1971). The fibroblasts were taken from a small biopsy of human skin treated enzymatically (Germain et al. 1993; Auger et al. 1995). The cells were characterized in culture to ensure their purity and their phenotype, then multiplied in vitro, in order to obtain enough cells for the reconstruction of each layer of the vascular wall: the intima (composed of EC), the media, and the adventitia. In order to obtain an abundant extracellular matrix production, fibroblasts and SMC were cultured in media supplemented with ascorbic acid until they self-assembled into sheet that could be detached from the culture support and then be wrapped around a tubular mandrel. According to this methodology, the steps necessary to produce a complete blood vessel were the following: the SMC sheet was rolled over an acellular inner membrane (dehydrated tubular tissue formed with a fibroblast sheet) to form the media. After 1 week of media maturation, a fibroblast sheet was wrapped around the media layer to produce the adventitia, which was kept in culture during 8 weeks for the maturation process. Finally, after eliminating the tubular mandrel, EC were seeded on the inner membrane by injection into the lumen to form a confluent endothelium. The macroscopic view of the TEBV is presented in Fig. 1.

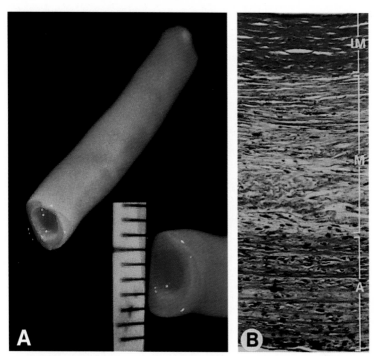

Fig. 1. Macroscopic (**A**) and microscopic (**B**) views of the mature tissue-engineered blood vessel (TEBV). **A** When removed from the tubular mandrel, the TEBV is self-supporting with an open lumen (3 mm internal diameter).
B Paraffin cross-section of the TEBV wall stained with Masson's trichrome. Collagen fibers are stained in *blue-green* and cells in *dark purple*. Inner membrane (*IM*)=125 µm, media (*M*)=320 µm and adventitia (*A*)=235 µm. Note the endothelium covering the luminal surface of the TEBV. Reprinted by permission from L'Heureux et al. (1998)

3.2 Mechanical, Histological, and Physiological Properties of the Tissue Engineered Blood Vessels

Although completely biological, this reconstructed blood vessel was highly resistant with a burst strength of over 2,500 mm Hg (Fig. 2B). This resistance is significantly higher than that of the human saphenous vein, considered to be the best biological material for lower-limb vascu-

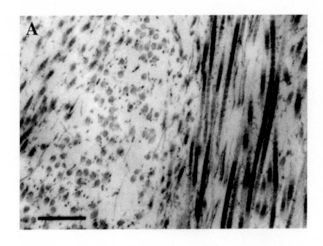

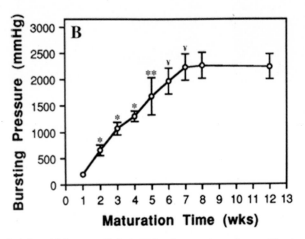

Fig. 2. A Adventitial extracellular matrix ultrastructure observed by transmission electron microscopy. Uranyl acetate and lead citrate staining (*bar*=500 nm). B Burst strength of the adventitia over time of maturation in vitro. *Significantly different from the precedent point (*$p<0.001$, **$p<0.005$, ¥$p<0.05$) with the Student's *t* test (*n*=8-13). Reprinted by permission from L'Heureux et al. (1998)

lar reconstruction (Veith et al. 1986; Abbott and Vignati 1995). Moreover, this resistance of the mature vessel was constant over time. This impressive resistance is attributed to the well-organized extracellular matrix composed of collagen fibrils that were oriented in directions perpendicular to one another in the concentric layers of the adventitia (Fig. 2A). Moreover, a constant gelatinase activity was measured starting at 2 weeks of adventitia maturation. Finally, our TEBV showed the presence of elastic fibers, a crucial component of the extracellular matrix sorely lacking in other reconstructed models (Weinberg and Bell 1986). All these critical characteristics were obtained even without pulsatile conditions of culture, which could, if applied, only improve the present model.

The media layer of the TEBV was evaluated for vasoreactivity to agonists and these elicited a contractile response to histamine in a similar fashion to the umbilical artery used as control (L'Heureux et al. 2001).

Histological and immunohistological analyses of the TEBV showed cells surrounded by a dense extracellular matrix composed of collagen and elastin. Interestingly, desmin, a protein component of the cellular intermediate filaments, known to be lost in culture (Thyberg et al. 1990; Christen et al. 1999), is re-expressed in SMC of the TEBV. This result indicates how close to its physiological counterpart our model is. Only quiescent SMC can produce such a molecule.

The EC seeded on the inner membrane formed a confluent monolayer (Fig. 1) expressing von Willebrand factor and the cells were able to incorporate acetylated low-density lipoproteins. This endothelium inhibited platelet adhesion in contrast to the non-endothelialized acellular membrane. Thus, this endothelium was functional and provided an anti-thrombotic surface (L'Heureux et al. 1998).

These human TEBV were implanted for 1 week in femoro-femoral interposition in dogs. Because of the xenogeneic situation, the prostheses did not contain EC. The implanted prostheses demonstrated that they could be easily handled and sutured by the use of conventional surgical techniques. A patency level of 50% was obtained after 1 week of implantation, and the patent implants were exempt of early tearing or dilatation (L'Heureux et al. 1998).

Thus, the TEBV we have produced offers exciting perspectives in clinical as well as in pharmacological applications. Currently, its production in vitro takes approximately 3 months, but this relatively long delay could be largely compensated for by the benefit of producing autologous implants for patients in possible future clinical applications; experiments are being conducted to ensure its long-term stability after implantation. Furthermore, appropriate culture conditions under pulsatile flow should decrease the production time of these TEBV.

4 Skin Equivalent Elaboration by Self-Assembly Approach

We recently fabricated the first skin equivalent composed exclusively of human cells into which pilosebaceous units were integrated (Michel et al. 1999). This could be the first step to more complex skin reconstruction by the self-assembly approach.

Fibroblasts, isolated from the dermal portion of skin biopsies and multiplied in vitro (Germain et al. 1993; Auger et al. 1995) were cultured for 35 days in medium supplemented with ascorbic acid in order to increase the production of extracellular matrix and form sheets of cells (Michel et al. 1999). The latter were then peeled from the culture support and four layers were superimposed. The construction was maintained by a stainless-steel anchoring ring for 1 week in order to allow further cell-matrix reorganization. Hair follicles were obtained by enzymatic treatment and forceps extraction from normal skin (removed during reductive breast surgery). The hair follicles were inserted in holes made in the multilayer dermal equivalent, and an additional fibroblast sheet was added underneath the hair follicles. One week later, keratinocytes were seeded and cultured until confluence occurred (1 week). Finally, the skin equivalent was maintained for 21 days at an air–liquid interface to induce the cornification of the epidermal layer (Pruniéras et al. 1983; Michel et al. 1999).

Photonic and transmission electron microscopy analyses revealed cell stratification (cuboidal in the basal layer with hemidesmosomes, cells becoming flat and elongated in the supra-basal layers) and the expression of typical markers of keratinocyte differentiation (keratin 10, filaggrin, and transglutaminase). The dermal portion displayed a remarkably "tissue-like" organization, with abundant collagen fibrils organized in parallel bundles perpendicular to one another and parallel to the epidermal surface like a normal dermis (Fig. 3C). One major and

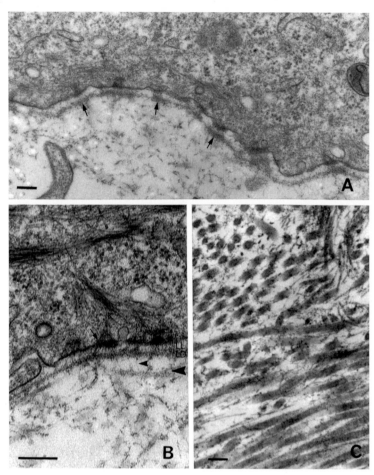

Fig. 3A–C. Transmission electron micrographs of the reconstructed skin architecture. **A** The dermal-epidermal junction contains hemidesmosomes (arrows) present all along the continuous basement membrane (*bar*=32 nm).
B Keratin intermediate filaments are attached to the dense plaque of the hemidesmosomes and span the lamina lucida for cell anchorage (*bar*=56 nm).
C The dermal extracellular matrix contains abundant network of collagen fiber bundles perpendicularly oriented and surrounded by a microfibrillar network (*bar*=32 nm). Reprinted by permission from Michel et al. (1999)

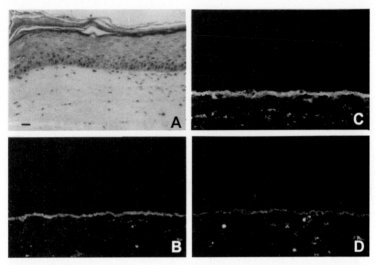

Fig. 4A–D. Histology and immunohistology of the basement membrane components of the skin equivalent. Cells are stained with hematoxylin, phloxine and saffron (**A**). Immunostaining with anti-laminin (**B**), anti-collagen type IV (**C**), and anti-collagen type VII (**D**) (*bar*=25 μm). Reprinted by permission from Michel et al. (1999)

interesting property displayed by our skin model was the development of the following highly differentiated structures: a continuous basal lamina, characterized by the expression of laminin and collagen types IV and VII (Fig. 4), and separated from the basal keratinocytes by a clear and distinct lamina lucida (Figs. 3A and 3B). We also observed the presence of numerous anchoring zones with keratin intermediate filaments attached to the dense plaque of the hemidesmosomes of basal keratinocytes (Figs. 3A and 3B). Moreover, the complete lamina densa was observed after only 4 weeks of co-culture, thus 1 week before the remarkable observations made by Contard et al. (1993). The hair follicle included in the model adapted its form to the insertion cavity and the follicle differentiation pattern was maintained: a basement membrane was detected at the dermal-follicular junction (laminin and type IV collagen production) as was a repartition of the filaggrin, keratin 10, and transglutaminase markers identical to that of a normal skin follicle.

Expression of trichohyalin, a hair-specific protein, was also detected (Michel et al. 1999).

This fully tissue-engineered skin equivalent demonstrated an excellent reorganization showing highly differentiated and rapidly formed structures similar to the original skin. Thus, the self-assembly approach allowed the in vitro reconstruction of a new histological-like model, which could be useful for the understanding of the mechanisms governing the interaction of cells and cell-matrix organization (Laplante et al. 2001). Furthermore, these substitutes may have distinct clinical advantages.

The self-assembly method for organ and tissue reconstruction is a truly novel approach in the field of tissue engineering of soft tissues (L'Heureux et al. 1998; Michel et al. 1999). The integration of various tissues and organs created by such a method should be more rapid and should lack the resorption phase that characterizes the scaffold approach in which synthetic or biosynthetic material has to be degraded by the implanted patient. This has lead to a chronic inflammatory reaction in many instances. By contrast, the self-assembly approach is akin to the physiological phenomenon that occurs in the womb during organogenesis. This may prove to be a fascinating advantage for certain types of grafts.

Furthermore, our research team has used several tissue-engineering approaches with success for the reconstruction of skin (Auger et al. 1995; Michel et al. 1995, 1999; Berthod et al. 1997; Black et al. 1998), blood vessels (L'Heureux et al. 1993, 1998), ligaments (Goulet et al. 1997a,b), bronchi (Paquette et al. 1998), and cornea (Germain et al. 1999, 2000).

These advances in tissue engineering will help in numerous areas of research on physiological, pathophysiological, and pharmacological mechanisms, permitting a better comprehension of these processes and the emergence of highly original methodologies to produce tissue-engineered implants, which could then solve the dramatic problem of organ supply.

References

Abbott WM, Vignati JJ (1995) Prosthetic grafts: when are they a reasonable alternative? Semin Vasc Surg 8:236–245

Allaire E, Guettier C, Bruneval P, Plissonnier D, Michel JB (1994) Cell-free arterial grafts: morphologic characteristics of aortic isografts, allografts, and xenografts in rats. J Vasc Surg 19:446–456

Auger FA (1988) The role of cultured autologous human epithelium in large burn wound treatment. Transplantation\Implantation Today 5:21–26

Auger FA, López Valle CA, Guignard R, Tremblay N, Noël B, Goulet F, Germain L (1995) Skin equivalent produced with human collagen. In vitro Cell Dev Biol Anim 31:432–439

Auger FA, Rouabhia M, Goulet F, Berthod F, Moulin V, Germain L (1998) Tissue-engineered human skin substitutes developed from collagen-populated hydrated gels: clinical and fundamental applications. Med Biol Eng Comput 36:801–812

Badylak SF, Lantz GC, Coffey A, Geddes LA (1989) Small intestinal submucosa as a large diameter vascular graft in the dog. J Surg Res 47:74–80

Badylak SF, Kropp B, McPherson T, Liang H, Snyder PW (1998) Small intestinal submucosa: a rapidly resorbed bioscaffold for augmentation cystoplasty in a dog model. Tissue Eng 4:379–387

Bell E, Ivarsson B, Merrill C (1979) Production of a tissue-like structure by contraction of collagen lattices by human fibroblasts of different proliferative potential *in vitro*. Proc Natl Acad Sci USA 76:1274–1278

Berthod F, Hayek D, Damour O, Collombel C (1993) Collagen synthesis by human fibroblasts cultured within a collagen sponge. Biomaterials 14:749–754

Berthod F, Germain L, Guignard R, Lethias C, Garrone R, Damour O, van der Rest M, Auger FA (1997) Differential expression of collagens XII and XIV in skin and in reconstructed skin. J Invest Dermatol 108:737–742

Black AF, Berthod F, L'Heureux N, Germain L, Auger FA (1998) In vitro reconstruction of a human capillary-like network in a tissue-engineered skin equivalent. FASEB J 12:1331–1340

Boyce ST, Christianson DJ, Hansbrough JF (1988) Structure of a collagen-GAG dermal skin substitute optimized for cultured human epidermal keratinocytes. J Biomed Mater Res 22:939–957

Boyce ST, Michel S, Reichert U, Shroot B, Schmidt R (1990) Reconstructed skin from cultured human keratinocytes and fibroblasts on a collagen-glycosaminoglycan biopolymer substrate. Skin Pharmacol 3:136–143

Brewster DC, Rutherford RB (1995) Prosthetic grafts in vacsular surgery. Saunders, Philadelphia, pp 492–521

Charara J, Beaudoin G, Fortin C, Guidoin R, Roy PE, Marble A, Schmitter R, Paynter R (1989) In vivo biostability of four types of arterial grafts with impervious walls: their haemodynamic and pathological characteristics. J Biomed Eng 11:416–428

Christen T, Bochaton-Piallat ML, Neuville P, Rensen S, Redard M, van Eys G, Gabbiani G (1999) Cultured porcine coronary artery smooth muscle cells. A new model with advanced differentiation. Circ Res 85:99–107

Contard P, Bartel RL, Jacobs L 2d, Perlish JS, MacDonald ED 2d, Handler L, Cone D, Fleischmajer R (1993) Culturing keratinocytes and fibroblasts in a three-dimensional mesh results in epidermal differentiation and formation of a basal lamina-anchoring zone. J Invest Dermatol 100:35–39

Cooper ML, Hansbrough JF, Spielvogel RL, Cohen R, Bartel RL, Naughton G (1991) In vivo optimization of a living dermal substitute employing cultured human fibroblasts on a biodegradable polyglycolic acid or polyglactin mesh. Biomaterials 12:243–248

Damour O, Braye F, Foyatier JL, Fabreguette A, Rousselle P, Vissac S, Petit P (1997) Cultured autologous epidermis for massive burn wounds: 15 years of practice. In: Rouabhia M (ed) Skin substitute production by tissue engineering: clinical and fundamental applications. Landes, Austin, pp 23–45

Dardik H (1989) Modified human umbilical vein allograft. In: Brutherford (ed) Vascular surgery. Saunders, Philadelphia, pp 474–480

Duplan-Perrat F, Damour O, Montrocher C, Peyrol S, Grenier G, Jacob M-P, Braye F (2000) Keratinocytes influence the maturation and organization of the elastin network in a skin equivalent. J Invest Dermatol 114:365–370

Gallico GG III, O'Connor NE, Compton CC, Kehinde O, Green H (1984) Permanent coverage of large burn wounds with autologous cultured human epithelium. N Engl J Med 331:448–451

Germain L, Rouabhia M, Guignard R, Carrier L, Bouvard V, Auger FA (1993) Improvement of human keratinocyte isolation and culture using thermolysin. Burns 2:99–104

Germain L, Auger FA, Grandbois E, Guignard R, Giasson M, Boisjoly H, Guérin SL (1999) Reconstructed human cornea produced in vitro by tissue engineering. Pathobiology 67:140–147

Germain L, Carrier P, Auger FA, Salesse C, Guérin S (2000) Can we produce a human corneal equivalent by tissue engineering? Prog Retinal Eye Res 19(5):497–527

Goulet F, Germain L, Caron C, Rancourt D, Normand A, Auger FA (1997a) Tissue-engineered ligament. In: Yahia LH (ed) Ligaments and ligamentoplasties. Springer, Berlin Heidelberg New York, pp 367–377

Goulet F, Germain L, Rancourt D, Caron C, Normand A, Auger FA (1997b) Tendons and ligaments. In: Lanza R, Langer R, Chick WL (eds) Principles of tissue engineering. Landes / Academic, Austin, Texas, pp 633–644

Heimbach D, Luterman A, Burke J, Cram A, Herndon D, Hunt J, Jordan M, McManus W, Solem L, Warden G, Zawacki B (1988) Artificial dermis for major burns. A multi-center randomized clinical trial. Ann Surg 208:313–320

Hirai J, Matsuda T (1996) Venous reconstruction using hybrid vascular tissue composed of vascular cells and collagen: tissue regeneration process. Cell Transplant 5:93–105

Huynh T, Abraham G, Murray J, Brockbank K, Hagen PO, Sullivan S (1999) Remodeling of an acellular collagen graft into a physiologically responsive neovessel. Nat Biotechnol 17:1083–1086

Isogai N, Landis W, Kim TH, Gerstenfeld LC, Upton J, Vacanti JP (1999) Formation of phalanges and small joints by tissue-engineering. J Bone Joint Surg Am 81:306–316

Jaffe EA, Nachman RL, Becker CG, Minick CR (1973) Culture of human endothelial cells derived from umbilical veins. Identification by morphologic and immunologic criteria. J Clin Invest 52:2745–2756

Jaksic T, Burke JF (1987) The use of "artificial skin" for burns. Annu Rev Med 38:107–117

Laplante AF, Germain L, Auger FA, Moulin V (2001) Mechanisms of wound reepithelialization: hints from a tissue-engineered reconstructed skin to long-standing questions. FASEB J (in press)

L'Heureux N, Germain L, Labbé R, Auger FA (1993) *In vitro* construction of human vessel from cultured vascular cells: a morphologic study. J Vasc Surg 17:499–509

L'Heureux N, Pâquet S, Labbé R, Germain L, Auger FA (1998) A completely biological tissue-engineered human blood vessel. FASEB J 12:47–56

L'Heureux N, Stoclet JC, Auger FA, Lagaud GJ-L, Germain L, Andriantsitohaina R (2001) A human tissue-engineered vascular media: a new model for pharmacological studies of contractile responses. FASEB J 15:515–524

Lin VS, Lee MC, O'Neal S, McKean J, Sung KL (1999) Ligament tissue engineering using synthetic biodegradable fiber scaffolds. Tissue Eng 5:443–452

López-Valle CA, Auger FA, Rompré P, Bouvard V, Germain L (1992) Peripheral anchorage of dermal equivalents. Br J Dermatol 127:365–371

Michel M, Germain L, Bélanger PM, Auger FA (1995) Functional evaluation of anchored skin equivalent cultured in vitro: percutaneous absorption studies and lipid analysis. Pharm Res 12:455–458

Michel M, L'Heureux N, Pouliot R, Xu W, Auger FA, Germain L (1999) Characterization of a new tissue-engineered human skin equivalent with hair. In vitro Cell Dev Biol Anim 35:318–326

Niklason LE, Gao J, Abbott WM, Hirschi KK, Houser S, Marini R, Langer R (1999) Functional arteries grown in vitro. Science 284:489–493

O'Connor NE, Mulliken JB, Banks-Schlegel S, Kehinde O, Green H (1981) Grafting of burns with cultured epithelium prepared from autologous epidermal cells. Lancet 1:75–78

O'Donnell TF, Mackey W, McCullough JL, Maxwell SL, Farber SP, Deterling RA, Callow AD (1984) Correlation of operative findings with angiographic and noninvasive hemodynamic factors associated with failure of polytetrafluoroethylene grafts. J Vasc Surg 1:136–148

Paquette JS, Goulet F, Boulet LP, Laviolette M, Tremblay N, Chakir J, Germain L, Auger FA (1998) Three-dimensional production of bronchi in vitro. Can Respir J 5:43

Peter SJ, Miller MJ, Yasko AW, Yaszemski MJ, Mikos AG (1998) Polymer concepts in tissue engineering. J Biomed Mater Res 43:422–427

Pruniéras M, Régnier M, Woodley D (1983) Methods of cultivation of keratinocytes with an air-liquid interface. J Invest Dermatol 81:28s–33s

Ross R (1971) The smooth muscle cell. II. Growth of smooth muscle in culture and formation of elastic fibers. J Cell Biol 50:172–186

Sayers RD, Raptis S, Berce M, Miller JH (1998) Long-term results of femorotibial bypass with vein or polytetrafluoroethylene. Br J Surg 85:934–938

Shinoka T, Shum-Tim D, Ma PX, Tanel RE, Langer R, Vacanti JP, Mayer JE Jr (1997) Tissue-engineered heart valve leaflets: does cell origin affect outcome? Circulation 96:II102–II107

Stanke F, Riebel D, Carmine S, Cracowski JL, Caron F, Magne JL, Egelhoffer H, Bessard G, Devillier P (1998) Functional assessment of human femoral arteries after cryopreservation. J Vasc Surg 28:273–283

Stephen M, Loewenthal, J, Little JM, May J, Sheil AG (1977) Autogenous veins and velour Dacron in femoropopliteal arterial bypass. Surgery 81:314–318

Thyberg J, Hedin U, Sjolund M, Palmberg L, Bottger BA (1990) Regulation of differentiated properties and proliferation of arterial smooth muscle cells. Arteriosclerosis 10:966–990

Tranquillo RT, Girton TS, Bromberek BA, Triebes TG, Mooradian DL (1996) Magnetically orientated tissue-equivalent tubes: application to a circumferentially orientated media-equivalent. Biomaterials 17:349–357

Veith FJ, Gupta SK, Ascer E, White-Flores S, Samson RH, Scher LA, Towne JB, Bernhard VM, Bonier P, Flinn WR, Astelford P, Yao JST, Bergan JJ (1986) Six-year prospective multicenter randomized comparison of autologous saphenous vein and expanded polytetrafluoroethylene grafts in infrainguinal arterial reconstructions. J Vasc Surg 3:104–114

Weinberg CB, Bell E (1986) A blood vessel model constructed from collagen and cultured vascular cells. Science 231:397–400

Yannas IV, Burke JF, Orgill DP, Skrabut EM (1982) Wound tissue can utilize a
    polymeric template to synthesize a functional extension of skin. Science
    215:174–176
Ziegler T, Alexander RW, Nerem RM (1995) An endothelial cell-smooth mus-
    cle cell co-culture model for use in the investigation of flow effects on vas-
    cular biology. Ann Biomed Eng 23:216–225

7 Flat Membrane Bioreactor for the Replacement of Liver Functions

L. De Bartolo, A. Bader

1 Relevance of Liver Support

Each year in the U.S., approximately 150,000 people are hospitalised with liver disease and over 43,000 people die from it (Langer and Vacanti 1993). These numbers are expected to increase as the 4 million people currently infected with Hepatitis C advance to liver failure. Transplantation, the only effective means of treating liver failure, is not an option for many patients. Some are simply not sick enough to justify the massive cost, invasiveness and risk of a transplant, leaving them unaided today. Other patients are too sick to qualify, while others die awaiting a transplant.

Ironically, the liver is a highly regenerative organ (Michapoulos 1990). Some patients currently undergoing liver transplantation would not need this major surgery if there were a simpler means of obtaining liver function until their own organ had recovered. Over the past 30 years, various supportive therapies for patients with acute liver failure have been proposed. Detoxification-based methodologies for liver support such as dialysis, haemofiltration and haemoperfusion have proven ineffective because physical methods are not sufficient for the management of severe biochemical disorders.

Unlike other organs (lung, kidney, heart) which have one primary function, the liver has multiple functions essential to maintain life, including carbohydrate metabolism, synthesis of proteins amino acid metabolism, urea synthesis, lipid metabolism, drug biotransformation and waste removal.

To address the critical medical needs of liver-compromised patients, the development of an extracorporeal liver-assist device, using isolated liver cells, to which patients would be temporarily connected until they recovered or received a liver transplant, could be a promising approach. Components of the patient's blood are to be passed through the device, processed by living liver cells within the device and then returned to the patient using a dialysis-type procedure.

Since fulminant liver failure is potentially reversible, the extracorporeal bridging of liver function would also be beneficial until the patient's own liver resumed functional activity.

2 Development of Bioartificial Liver

Recent developments in tissue engineering have made it possible for us to use isolated hepatocytes in a bioreactor for the creation of a bioartificial liver, which supports patients with acute liver failure. Isolated hepatocytes retain tissue-specific functions and may able to correct metabolic imbalances while providing specific factors for liver regeneration. Since fulminant liver failure is potentially reversible, extracorporeal bridging of liver function would also be beneficial until the patient's own liver resumed functional activity. In recent years, different bioreactor systems have been developed. They can be classified according to the immobilisation technique used: microcarriers (Demetriou et al. 1986), hollow-fibre membranes (Gerlach et al. 1994), flat-sheet membranes (Bader et al. 1998), biological matrices (Bucher et al. 1990), non-woven polyester matrix (Flendrig et al. 1997) and encapsulation

(Dixit et al. 1990) amongst others. These bioreactors vary greatly with respect to microenvironment; the means of oxygen and nutrient supply to the cells and configuration and type of devices differ from one design to the other. These reports suggested that a bioartificial liver using isolated hepatocytes is a promising approach for treatment of patients with hepatic failure, although further studies to address the problems related to the development of a bioartificial liver are necessary. In fact, there is not still an ideal device. This is not surprising, given the complexity of liver functions. A liver support system must supply various liver-specific functions including synthetic and detoxification activities for a time sufficient to allow recovery of a patient or maintenance of patients until transplantation. It is increasingly evident that a multifactorial approach to liver support will be necessary.

In membrane bioartificial liver, semi-permeable membranes play more functions: they act as immunoselective barriers and as a means for cell oxygenation, and they provide a large area for cell attachment (De Bartolo and Drioli 1998). All these functions are important for the maintenance of cell viability and specific functions. In a membrane bioartificial liver, cells come into contact with the membrane surface. Therefore, the response of the cell behaviour depends on surface properties of the used membrane. For this reason, membranes should be chosen not only based on their separation properties but also on physicochemical and morphological surface properties (De Bartolo et al. 1999).

When designing an extracorporeal hybrid liver support device, special attention should be paid to providing an architectural basis for reconstructing a proper cellular microenvironment that ensures the highest and prolonged functional activity of liver cells cultured in the bioreactor. It should achieve high cell-density culture and the bioreactor should be designed at a full-scale under sufficient mass transfer conditions. Tissue architecture and parenchymal cell morphology are crucial for the proper functioning of liver cells within the organ in situ. In the liver, hepatocytes are organised three-dimensionally into cellular plates that are identified as cords in a cross-sectional view. In these plates, hepatocytes exhibit a distinctive epithelial polarity, as well as a strongly developed cell–cell communication structures including bile canaliculi and tight junctions. Thus, hepatocytes must interact with other cells as well as with chemically complex substrata to sustain viability and functions. This organisation allows them to obtain an adequate supply of oxygen and nutrients.

It has been well demonstrated that the extracellular matrix induces both cell polarity and tissue organisation. In fact, it was reported that the culture of hepatocytes within the extracellular matrix in a 3-D system amplifies their specific metabolic activities and cell stability; hepatocytes in the extracellular matrix culture maintained a stable level of albumin and stable transcription and secretory rates of proteins (Bader et al. 1992; Dunn et al. 1992). A key issue in the development of a bioartificial liver is the development in vitro of a 3-D culture system which imitates more closely the in vivo organisation.

3 Hepatocyte Source for Bioartificial Liver

The ideal type of hepatocytes to be used for bioartificial liver are human normal hepatocytes readily available, at high density, stable for weeks and able to replace synthetic and detoxification functions of liver. However, there is a shortage of cadaveric organs. Tumour-derived cell lines have been used as alternative but it is known that tumour cell lines shows alteration of gene expression under culture conditions and may exhibit lower liver-specific functions than primary cultured cells (Nyberg et al. 1994). Furthermore, one of the major problems with this approach is the potential leakage of tumour cells and their products into patient's circulation. A much safer approach is the use of immortalised but non-tumorigenic human hepatocytes with the same function phenotype as primary cells (Liu et al. 1996). Adult hepatocytes do not proliferate in culture. Hepatocytes isolated from embryonic/neonatal animal liver or from hepatectomised adult are able to proliferate. However, the use of this type of hepatocytes is not convenient because they show incomplete expression of differentiated functions. Finally, epithelial growth factors and hepatic growth factors were identified as hepatocyte mitogens, so that it is also possible to stimulate the proliferation of primary cells in vitro. As a result, porcine hepatocytes are probably one of the best xenograft candidates with regard to differentiated metabolic functions and high-yield obtention. Pig cells have similar metabolism to human cells. One pig liver may provide cell mass sufficient for several devices.

4 Bioartificial Liver: Our Design

Our goal was to evaluate the biochemical performance of a full-scale flat membrane bioreactor (FMB) developed by Bader et al. (1998). The FMB permits high-density hepatocyte culture and simultaneously allows us to culture cells under sufficient oxygenation conditions as in vivo-like microenvironment. In such a bioreactor, which was built according to the in vivo organisation, pig liver cells are organised in a simple and repetitive way: hepatocytes are arranged as a plate in 3-D co-culture with non-parenchymal cells within extracellular matrix between oxygen-permeable flat-sheet membranes. The membranes permit unlimited oxygen uptake by the hepatocytes and the right geometry for the attachment of hepatocytes as monolayer (Bader et al. 1999; De Bartolo et al. 1999). In hybrid liver support devices where oxygen is solely transported by diffusion, meeting the cell oxygen metabolic needs requires that the bioreactor ensures an oxygen path length shorter than about 100–150 µm (Catapano et al. 1996). In the FMB the mean diffusion distance from oxygen permeable membranes to cells is about 20 µm. Embedding of hepatocytes within two layers of extracellular matrix improves the maintenance of liver specific functions of cells (Bader et al. 1992; Dunn et al. 1992). Another important goal was to develop a bioreactor that could be scaled-up to incorporate sufficient cell mass for possible therapeutic liver support.

4.1 Features of Full-Scale Flat Membrane Bioreactor

The bioreactor consists of a multitude of stackable flat membrane modules, each having an oxygenating surface area of 1150 cm^2 (De Bartolo et al. 2000). Up to 50 modules can currently be run in parallel mode. Each module with 2×10^8 cells seeded results in a total seeded number of 1×10^{10} cells. We used up to 20 modules (4×10^9 cells). The hepatocytes were located at a distance of 10–20 µm of extracellular matrix. Medium and cells in the modules were oxygenated in incubator by diffusion of humidified air across non-porous polytetrafluoroethylene membrane (DuPont) with a 35-µm thickness. A microporous membrane with 0.28 µm mean pore size that separates the cell compartment from the medium compartment is generally used when the FMB is used at high

Fig. 1. Flat membrane bioreactor

flow rate in order to protect cells from shear forces and to control mass transfer of different solutes. The modules are separated each from the other and are connected to a medium reservoir with tubing by multi-channel peristaltic pump, which controls flow rate in the inlet and outlet stream (Fig. 1). In these experiments, the FMB was perfused with serum-free medium at a flow rate of 9 ml/h. The bioreactor was prepared for cell culture by coating of the oxygen permeable with rat-tail type I collagen. After seeding and attachment of the hepatocytes, a second layer of matrix was placed on top.

One of the key features of the bioreactor is its novel multi-modular design. Bioreactor modules can thus be individually observed under a microscope. Both sides of the outside shell consist of gas-permeable oxygen-permeable polytetrafluoroethylene (PTFE) membranes, which allow direct oxygenation of the cell plate. Porcine hepatocytes are maintained in 3-D co-culture with non-parenchymal cells. The cell plate is simultaneously enclosed within two layers of extracellular matrix. This is a further development of our previously reported 3-D co-culture model (Bader et al. 1998).

In order to calculate the rates of metabolic reactions we evaluated the fluid dynamics of the bioreactor with tracer experiments. The bioreactor without cells was challenged by changing the tracer concentration stepwise in the feed stream, and the outlet tracer concentration was established by assaying samples of the effluent collected at regular intervals. As tracer, we used trypan blue in water. The agreement of the experimental F(*t*) with that of a plug flow reactor was assessed by comparing experimental data with model of ideal plug flow reactor.

4.2 Scale-up of the Flat Membrane Bioreactor

Clinical treatment of hepatic failure requires high cell concentration inside of the bioreactor, which is often realised by formation of cell aggregates of large size. In this case, resistances to mass transport results in a depletion of oxygen and nutrients in the central cell regions farther away from the external surrounding cell layers, and hence in cell starvation and death. Generally in most of the hollow fibre bioreactors existing to date, the transport of nutrients and metabolites is realised by diffusion, which is known to be a limiting mechanism of mass transport. In the liver the problem of oxygen and nutrients supply to the cells is solved by arranging them in cell plates with sinusoidal structures located on both sides. In contrast to other bioreactors, our device is based on the organisation of liver cells as a plate within extracellular matrix in which each individual hepatocyte has its own membrane support and thereby its own oxygen supply position (see Fig. 1).

The characteristics of the bioreactor design is its modular structure that permits an increase in the number of stackable flat-membrane modules until the required cell mass has been obtained without changes of the individual module configuration. Often, bioreactor perform very well under laboratory-scale conditions and lose functions under clinical-scale conditions. For this reason, the fluid dynamics characterisation of the bioreactor in the scale-up process plays a key role. The fluid dynamics characterisation provided information about the actual dynamic behaviour of the designed bioreactor under operating conditions and independent on the type of cells. In this way, it was possible to culture liver cells under known fluid dynamics conditions and to make at steady state a reliable and effective estimation of cell metabolic reaction rate. The importance of a fluid dynamics characterisation is further stressed by the fact that, in the scale-up process from the laboratory to the clinical size, the deviation from an ideal fluid dynamics behaviour is often beyond the control of the designer and, for practical reasons, is generally different in small- and large-scale devices.

As can be deducted from Fig. 1, scaling-up the FMB implies increasing the number of stackable flat-membrane modules (De Bartolo et al. 2000). This does not affect the individual bioreactor configuration, since the dimensions and the thickness of the respective flat-membrane modules remain the same.

The full-scale bioreactor was used previously in static-condition culture as reported elsewhere and later in flux conditions (Bader et al. 2000; De Bartolo et al. 2000).

5 Pig Liver Cell Isolation and Culture in the Flat Membrane Bioreactor

Liver cells were isolated from female pigs weighing from 20 to 25 kg as described previously (De Bartolo and Bader 2000). Viability of the hepatocytes ranged between 90% and 95% as assessed by trypan blue exclusion. Freshly isolated hepatocytes were seeded in the FMB on gas-permeable surfaces pre-coated with rat-tail collagen type. Bioreactors without hepatocytes were used as controls. The perfusion of the FMB was initiated in single-pass 24 h after cell loading. Media samples were collected daily on the inlet and outlet stream to determine product formation and substrate clearance.

The morphology and the concentration of cells cultured inside the FMB were assessed by inverted light microscopy for 18 days of culture. The liver-specific functions of isolated hepatocytes were investigated by estimating their ability to synthesise albumin. Albumin secretion was measured in the samples by an immunometric method (enzyme-linked immunosorbent assay). The ability of the bioreactor to perform drug biotransformation functions was evaluated by investigating diazepam metabolism. Diazepam metabolism by hepatocytes was assessed by estimating the diazepam elimination and the formation of its metabolites in the presence of 10 µg/ml diazepam in the culture medium. Diazepam and metabolite concentrations were assessed by analysis of HPLC (Merck/Hitachi, Darmstadt, Germany).

Experimental metabolic rates are reported as mean±standard error of the mean of six replicated cultures for each time point from three different isolations. The statistical significance of the experimental results was established according to the Student's t test.

6 Performance of the Full-Scale Flat Membrane Bioreactor

6.1 Morphology and Viability of Cells Inside the FMB

The design of the bioreactor permitted the on line observation of the cells with an inverse microscope. Porcine hepatocytes were isolated as small aggregates. A small fraction of non-parenchymal cells was also obtained. When loaded in the FMB, isolated cells rapidly attached and formed small clusters. They spread within 24 h and re-established intercellular contacts with near cells. During the first days of culture, cells proliferated and reached a completely confluent monolayer at 5 days in

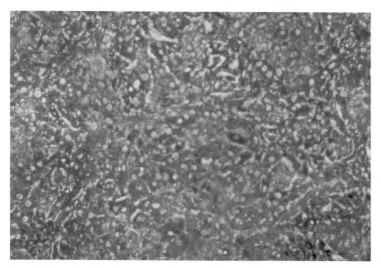

Fig. 2. Inverse light microscopy of isolated hepatocytes cultured inside the FMB after 8 days of culture

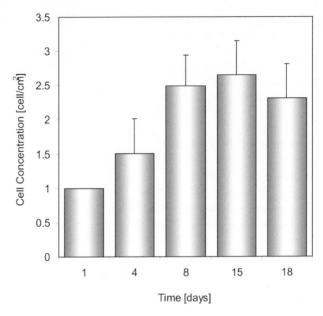

Fig. 3. Cell concentration inside the flat membrane bioreactor

culture. Cultures were typically characterised by polyhedral cell shapes and clearly distinguishable intercellular biliary zones for all investigated period (Fig. 2).

The morphological observation with regard to the initial increase of cell number on the surface with time is confirmed by measurement of cell concentration in the FMB. Figure 3 shows the changing of cell concentration in the FMB with culture time: cell concentration increased during the time of culture and peaked at day 8, reaching a 2.5-fold value compared with cells at 1 day of culture. Thereafter, the number of cells remained at a concentration of about 2 cells/cm^2 until about 3 weeks of culture.

The potential of liver cells to reconstitute many of the features as in vivo and to increase their cell concentration inside of the bioreactor is probably due to co-culture of hepatocytes with non-parenchymal cells. Indeed, it has been shown that the survival and functions of hepatocytes improved when they were co-cultured with non-parenchymal cells

(Bader et al. 1996). The use of a flat sheet device is especially advantageous when hepatocytes are intended to be cultured in a cell-plate configuration, as done in this study using a 3-D co-culture model. This is a further development of the sandwich model and our previous co-culture model, which placed the non-parenchymal cells on top of the second layer of matrix. In the current study we have, on the contrary, included the non-parenchymal cells within the hepatocellular layer.

6.2 Liver Synthetic and Drug Biotransformation Functions of the FMB

Various functions of hepatocytes in vivo are regulated by cell–cell contact, and the stable functions are not maintained unless the cells from tissue receive complex regulation from their microenvironment (Nakamura et al. 1983). Specific metabolic functions of hepatocytes in terms of albumin synthesis are sustained for the investigated culture time demonstrating thus the long-term maintenance of functional integrity of hepatocytes cultured in the FMB. To evaluate the performance of FMB, each bioreactor was monitored for 18 days for albumin synthesis. Figure 4 shows that the ability of hepatocytes to synthesise albumin increased during the first days of culture. High rates of albumin synthesis were obtained at day 10. Thereafter, rates of albumin production remained at a value of 1.41 pg/h/cell until 18 days of culture. However, at that time, the rate of albumin production was significantly higher than that exhibited by cells during the first days of culture.

Diazepam was used to investigate the capacity of the FMB to perform drug biotransformation functions. Furthermore, it seems that endogenous benzodiazepine analogues play a pivotal role in the pathogenesis of hepatic encephalopathy; a relationship between plasma benzodiazepine receptor ligand concentrations and severity of hepatic encephalopathy associated with acute liver failure has been shown (Baker et al. 1990).

Diazepam clearance and metabolite formation were investigated in samples obtained from culture. Diazepam biotransformation led to the formation of three metabolites, typically found also in vivo in humans. Diazepam is metabolised through *N*-demethylation and C3-hydroxylation to *N*-desmethyldiazepam (nordiazepam) and 3-hydroxydiazepam

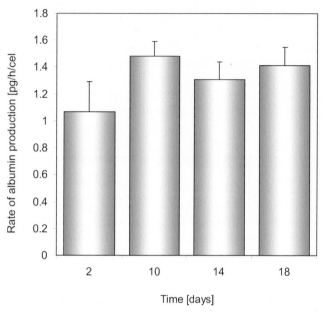

Fig. 4. Rate of albumin production of porcine hepatocytes cultured in FMB for 18 days (enzyme-linked immunosorbent assay [ELISA], average±standard deviation of six experiments)

(temazepam) respectively, with either or both being converted to oxazepam. The metabolite formation in the bioreactor is equally composed of the same three different metabolites. Porcine hepatocytes cultured in the FMB were able to degrade the diazepam present in the culture medium; from day 2 to day 11 the diazepam concentration added to the culture medium decreased significantly reaching values which are from 4.3 to 6.8 times lower than those present in the control. As shown in Fig. 5, the production of diazepam metabolites nordiazepam and temazepam did not significantly change over a period of 18 days but remained more or less at the same value. Correspondingly, the subsequent rates of conversion of nordiazepam and temazepam to oxazepam were constant during the culture time, with values 0.58 and 0.62 ng/h/cell.

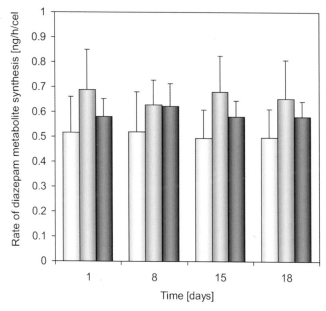

Fig. 5. Formation of diazepam metabolites by porcine hepatocytes cultured in the FMB in the presence of 10 μg/ml diazepam. *White bars*, rate of temazepam synthesis; *grey bars*, rate of nordiazepam synthesis; *black bars*, rate of oxazepam synthesis. (HPLC; average±standard deviation of six experiments)

Hepatic encephalopathy is possibly caused by an interaction of endogenous benzodiazepines and elevated ammonia levels. Basile et al. (1991) detected diazepam in patients suffering from acute liver failure, hepatic encephalopathy without previous exposure to diazepam. In our study, we used the application of exogenous diazepam at high levels as a tool to investigate the potential of the newly developed flat membrane bioreactor to cope with this drug. The results obtained clearly demonstrate the capacity of the bioreactor system to detoxify diazepam. Diazepam is metabolised by cytochrome P450 activities. These enzymes are among the most sensitive and fragile found in hepatocytes, responding quickly by loss of activities to unfavourable culture conditions. The high cell-specific activity found in this study is a clear proof for the workability of the device concept.

7 Conclusions

These results demonstrate that full-scale FMB sustained viability, liver-specific functions and drug biotransformation functions of hepatocytes over the prolonged culture period. The investigated metabolic reactions for hepatocytes described here are comparable with those reported by other authors The scale-up of the FMB was assessed, and it could be demonstrated that the device design aimed at the reconstruction of the liver-specific tissue architecture supports the expression of tissue specific functions. These characteristics encourage us to perform in vivo experiments in animal models in order to evaluate the potential of the FMB as bioartificial liver.

References

Bader A, Rinkes IHB, Closs IE, Ryan CM, Toner M, Cunningham JM, Tompkins GR, Yarmush ML (1992) A stable long-term hepatocyte culture system for studies of physiologic processes: cytokine stimulation of acute phase response in rat and human hepatocytes. Biotechnol Prog 8:219–225

Bader A, Knop E, Kern A, Boker K, Fruhauf N, Crome O, Esselmann H, Pape C, Kempka G, Sewing K-Fr (1996) 3-D coculture of hepatic sinusoidal cells with primary hepatocytes design of an organotypical model. Exp Cell Res 226:223–233

Bader A, De Bartolo L, Haverich A (1997) Initial evaluation of the performance of a scaled-up flat membrane bioreactor (FMB) with pig liver cells. In: Crepaldi G, Demetriou AA, Muraca M (eds) Bioartificial liver: the critical issues. CIC International Editions, Rome, pp 36–41

Bader A, Frühauf N, Zech K, Haverich A, Borlak J (1998) Development of a small scale bioreactor for drug metabolism studies maintaining hepatospecific functions. Xenobiotica 28:815–825

Bader A, Fruhauf N, Tiedge M, Drinkgen M, De Bartolo L, Borlak JT, Steinhoff G, Haverich A (1999) Enhanced oxygen delivery reverses anaerobic metabolic states in prolonged sandwich rat hepatocytes. Exp Cell Res 246(1):221–232

Bader A, De Bartolo L, Haverich A (2000) High level benzodiazepine and ammonia clearance by flat membrane bioreactors with porcine liver cells. J Biotechnol 81(2/3):95–105

Baker BL, Morrow AL, Vergalla J, Paul SM, Jones EA (1990) Gamma-aminobutyric acid (GABAA) receptor-function in a rat model of hepatic encephalopathy. Metab Brain Dis 5:185–193

Basile AS, Hughes R, Harrison PM, Murata Y, Pannell L, Jones EA, Williams R, Skolnick P (1991) Elevated brain concentrations of 1,4-benzodiazepines in fulminant hepatic failure. N Engl J Med 325:473–478

Bucher NL, Robinson GS, Farmer SR (1990) Effects of extra-cellular matrix on hepatocyte growth and gene expression: implications for hepatic regeneration and the repair of liver injury. Semin Liver Dis 10:11–19

Catapano G, De Bartolo L, Lombardi CP, Drioli E (1996) The effect of oxygen transport resistances on the viability and functions of isolated rat hepatocytes. Int J Artif Organs 19(1):61–71

De Bartolo L, Bader A (2000) Performance of flat membrane bioreactor utilizing porcine hepatocytes cultured in an extracellular matrix. In: Berry MN, Edwards AM (eds) The hepatocyte review. Kluwer Academic, Dordrecht, pp 525–534

De Bartolo L, Drioli E (1998) Membranes in artificial organs. In: Haris PI, Chapman D (eds) New biomedical materials – basic and applied studies. IOS Press, Amsterdam, pp 167–181 (Biomedical and health research, vol 16)

De Bartolo L, Catapano G, Della Volpe C, Drioli E (1999) The effect of surface roughness of microporous membranes on the kinetics of oxygen consumption and ammonia elimination by adherent hepatocytes. J Biomat Sci (Polymer Edn) 10(6):641–655

De Bartolo L, Jarosch-Von Schweder G, Haverich A, Bader A (2000) A novel full-scale flat membrane bioreactor utilizing porcine hepatocytes: cell viability and tissue specific functions. Biotechnol Prog 16(1):102–108

Demetriou AA, Whiting JF, Levenson AM, Chowdhury NR, Schechner R, Michalski S, Feldman D, Chowdhury JR (1986) New method of hepatocyte transplantation and extracorporeal liver support. Ann Surg 204(3):259–270

Dixit V, Darvasi R, Arthur M, Brezina M, Lewin K, Gitnick G (1990) Restoration of liver function in Gnunn rats without immunosuppression using transplanted microencapsulated hepatocytes. Hepatology 15:1342–1349

Dunn JCY, Tompkins RG, Yarmush ML (1992) Hepatocyes collagen gel sandwich: evidence for transcriptional and translational regulation. J Cell Biol 116:1043–1053

Flendrig LM, La Soe JW, Joerning GGA, Steenbeck A, Karlsen OT, Bovée WMM, Ladiges NCJJ, te Velde AA, Chamuleau AFM (1997) In vitro evaluation of a novel bioreactor based on an integral oxygenator and a spirally wound nonwoven polyestermatrix for hepatocyte culture as small aggregates. J Hepatol 26:1379–1392

Gerlach J, Schnoy N, Smith MD, Neuhaus P (1994) Hepatocyte culture between woven capillary network-A microscopy study. Artif Organs 18(3):1–5

Langer R, Vacanti P (1993) Tissue engineering. Science 260:920

Liu J, Naik S, Santangini H, Trenkler D, Chowdhury JR, Jauregui HO (1996) Characterisation of immortalised porcine hepatocytes. Hepatology 24(4):197A, 283

Michapoulos GK (1990) Liver regeneration: molecular mechanisms of growth control. FASEB J 4:176–187

Nakamura T, Yoshimoto K, Nakayama Y, Tomita Y, Ichihara A (1983) Reciprocal modulation of growth and differentiated functions of mature rat hepatocytes in primary culture by cell-cell contact and cell membranes. Proc Natl Acad Sci 80:7229–7233

Nyberg SL, Remmel RP, Mann HJ, Peshwa MV, Hu W, Verra FB (1994) Primary outperform hep G2 cells as source of biotransformation functions in a bioartificial liver. Ann Surg 220:59–67

8 Tissue Engineering of Genitourinary Organs

J.J. Yoo, A. Atala

1 Introduction

Congenital abnormalities, cancer, trauma, infection, inflammation, iatrogenic injuries, and other conditions may lead to genitourinary organ damage or loss, requiring eventual reconstruction. Whenever there is a lack of native urological tissue, reconstruction may be performed with native non-urological tissues (skin, gastrointestinal segments or mucosa from multiple body sites), homologous tissues (cadaver fascia, cadaver or donor kidney), heterologous tissues (bovine collagen), or artificial materials (silicone, polyurethane, Teflon).

Tissue engineering follows the principles of cell transplantation, materials science, and engineering towards the development of biological substitutes which would restore and maintain normal function. Tissue engineering may involve matrices alone, wherein the body's natural ability to regenerate is used to orient or direct new tissue growth, or the use of matrices with cells. When cells are used, donor tissue is dissociated into individual cells which are either implanted directly into the host, or expanded in culture, attached to a support matrix and re-implanted after expansion. The implanted tissue can be either heterologous, allogeneic, or autologous. Ideally, these approaches might allow lost tissue function to be restored or replaced in toto and with limited complications (Atala 1997). The approach which we have followed to bioengineer tissues involves the use of matrices alone and the use of such matrices with the eventual use of autologous cells – wherein a biopsy of tissue is obtained from the host, the cells are dissociated and expanded in vitro, reattached to a matrix, and implanted into the same host – thus avoiding rejection (Atala et al. 1992, 1993a,b, 1994; Cilento et al. 1994; Atala 1995, 1997, 1998, 1999; Yoo and Atala 1997; Fauza et al. 1998; Yoo et al. 1998a,b, 1999; Machluf and Atala 1998; Amiel and Atala 1999; Kershen and Atala 1999; Oberpenning et al. 1999; Park et al. 1999).

One of the initial limitations of applying tissue engineering techniques to urological organs had been the previously encountered inherent difficulty of growing genitourinary-associated cells in large quantities. In the past, it was believed that urothelial cells had a natural senescence which was hard to overcome. Normal urothelial cells could be grown in the laboratory setting but with limited expansion. A system of urothelial cell harvest was created in our laboratory which does not use any enzymes or serum and has a large expansion potential. Using these methods of cell culture, it is possible to expand a urothelial strain from a single specimen, which initially covers a surface area of 1 cm^2, to one covering a surface area of 4,202 m^2 (the equivalent area of one football field) within 8 weeks (Cilento et al. 1994). These studies indicated that it should be possible to collect autologous urothelial cells from human patients, expand them in culture, and return them to the human donor in sufficient quantities for reconstructive purposes. Bladder, ureter, and renal pelvis cells can be equally harvested, cultured, and expanded in a similar fashion. We have shown that normal human bladder epithelial and muscle cells can be efficiently harvested from surgical material, extensively expanded in culture, and their differentiation characteristics, growth requirements, and other biological properties studied (Folkman and Hochberg 1973; Cilento et al. 1994; Yoo et al. 1995, 1998; Cilento et al. 1996; Yoo and Atala 1997; Fauza et al. 1998; Oberpenning et al. 1999).

Cells or tissue cannot be implanted in volumes greater than 3 mm^3 (Folkman and Hochberg 1973), Nutrition and gas exchange is limited by this maximum diffusion distance. If cells were implanted in volumes greater than 3mm^3, only the cells on the surface would survive, and the central cell core would necrose, due to a lack of vascularity (Folkman and Hochberg 1973). How does nature solve the problem of maximum diffusion distances through tissues? Within all living organisms, branching is the solution to nutrient diffusion. In order to successfully transplant cells in large volumes, it may be necessary to imitate the branching pattern required for cell implantation and subsequent survival. The matrices which we have used are designed with a branching pattern which allows for cell attachment to the "branches" or "pores," thus allowing for capillary infiltration to the interstitial spaces after implantation in vivo (Filkman and Hochberg 1973; Atala et al. 1992, 1993; Yoo et al. 1995, 1998a,b, 1999; Cilento et al. 1996; Yoo and Atala 1997; Fauza et al. 1998; Atala 1999; Oberpenning et al. 1999; Park et al. 1999). In this manner, large numbers of cells can be implanted with maximal survival, or adjacent cells can migrate onto the matrix with its appropriate vascularity.

The success of using cell transplantation strategies for urological reconstruction depends on the ability to use donor tissue efficiently and to provide the right conditions for long-term survival, differentiation and growth. We have achieved an approach to tissue regeneration by patching isolated cells to a support structure which would have suitable surface chemistry for guiding the reorganization and growth of the cells. The supporting matrix is composed of biodegradable artificial or natural matrices which can allow cell survival by diffusion of nutrients across short distances once the cell-support matrix is implanted. The cell-support matrix becomes vascularized in concert with expansion of the cell mass following implantation.

Our laboratory has utilized both synthetic (polyglycolic acid polymer scaffolds alone and with co-polymers of poly-l-lactic acid and poly-DL-lactide-co-glycolide) and natural biodegradable materials (processed collagen derived from allogeneic donor bladder submucosa and intestinal submucosa) as cell delivery vehicles (Atala et al. 1992, 1993b, 1999; Yoo et al. 1995, 1998b, 1999; Cilento et al. 1996; Yoo and Atala 1997; Fauza et al. 1998; Atala 1999; Chen et al. 1999; Oberpenning et al. 1999; Park et al. 1999). These matrices have many desirable features; they are biocompatible and processable. Degradation occurs by hydrolysis, and the time sequence can be varied by changing the processing conditions. These biodegradable matrices, both natural and artificial, can be readily modified, depending on their intended application, into a variety of shapes and structures, including small diameter fibers and porous films, to various levels of rigidity or elasticity.

2 Tissue Engineering Ex Situ

The initial experiments were designed in order to explore the possibility of engineering urological tissue components ex situ (outside the urinary tract) in an in vivo animal model (Atala et al. 1992, 1993). Normal urothelial and muscle cells were expanded in vitro, seeded onto polyglycolic acid (PGA) polymer scaffolds either separately or together, and allowed to attach and form sheets of cells in vitro. Several parameters, regarding specific cell attachment properties on the polymers scaffolds, were evaluated over time. A series of in vivo cell–polymer scaffold experiments were performed. Normal primary bladder urothelial and muscle cells were expanded, seeded onto biodegradable polymer scaffolds, and implanted in the subcutaneous space of athymic mice, not in continuity with the urinary tract.

Histological analysis of normal primary urothelial, bladder muscle, and composite urothelial and bladder muscle–polymer scaffolds, implanted in athymic mice and retrieved at different time points, indicated that viable cells were evident in all three experimental groups. Implanted cells oriented themselves spatially along the polymer surfaces. The cell populations appeared to expand from one layer to several layers of thickness with progressive cell organization with extended implantation times. Polymers alone or with cells evoked an aggressive angiogenic response by 5 days, which increased with time. Polymer fiber degradation was evident after 20 days. An inflammatory response was

also evident at 5 days, and its resolution correlated with the biodegradation sequence of the polymers. Cell-polymer composite implants of
urothelial cells alone, retrieved at extended times (50 days), showed
extensive formation of multi-layered sheet-like structures (Atala et al.
1992). Polymers seeded with both muscle and urothelial cells and manipulated into a tubular configuration showed layers of muscle cells
lining the multilayered epithelial sheets (Atala et al. 1993b). Cell polymers implanted with human bladder muscle cells alone showed almost
complete replacement of the polymer scaffold with sheets of smooth
muscle at 50 days.

These experiments demonstrated that newly isolated human bladder
urothelial and muscle cells would attach to artificial and natural matrices
in vitro and that, when implanted into animals, these constructs could
survive, proliferate, and reorganize into newly formed multilayered
structures which exhibit spatial orientation and a normal histomorphol-
ogy in vivo (Atala et al. 1992, 1993b). These experiments also demonstrated, for the first time in the field of tissue engineering, that composite
tissue-engineered structures could be created de novo (Atala et al.
1993b). Prior to this study, only single-cell-type tissue-engineered structures had been created for any tissue or organ.

3 Bladder and Ureter

In order to determine the effects of implanting engineered tissues in
continuity with the urinary tract, experiments were performed in an
animal model of ureteral replacement. In one study conducted in dogs,
urothelial and smooth muscle cells were harvested, expanded in vitro,
and seeded onto biodegradable matrix scaffolds. These structures were
tubularized and used to replace ureteral segments in each animal (Yoo et
al. 1995). The malleability of the synthetic matrix allowed for the
creation of cell-matrix implants manipulated into pre-formed tubular
configurations. The combination of both smooth muscle and urothelial
cell-matrix scaffolds is able to provide a template wherein functional
ureteral tissue may be created de novo. In the studies performed, if an
entire segment was replaced, cells were needed in order to prevent graft
resorption and strictures. However, if the area replaced was small in at

least one of its dimensions, i.e., an onlay graft, the cells were not essential for adequate healing.

Bladder engineering experiments were initially performed using animal models of augmentation (Yoo et al. 1998a). Partial cystectomies were performed in dogs. Both urothelial and smooth muscle cells were harvested and expanded separately. A collagen matrix obtained from processed allogenic bladder submucosa was seeded with muscle cells on one side and urothelial cells on the opposite side. All dogs underwent cruciate cystotomies on the bladder dome. Augmentation cystoplasty was performed with either cell-seeded or unseeded matrices. Bladders augmented with the cell-seeded matrix scaffold showed a 99% increase in capacity compared to bladders augmented with the unseeded matrix, which showed only a 30% increase in capacity, wherein graft contraction and shrinkage occurred. Histologically, the retrieved engineered bladders contained a cellular organization consisting of a urothelial-lined lumen surrounded by submucosal tissue and smooth muscle. However, the muscular layer was markedly more prominent in the cell-reconstituted scaffold (Yoo et al. 1998a).

Most of the unseeded matrices utilized for bladder replacement in the past have been able to show adequate histology in terms of a well-developed urothelial layer; however, they have been associated with an abnormal muscular layer which varies in terms of its full development (Atala 1995, 1998). It has been well established for decades that the bladder is able to regenerate generously over free grafts. The urothelium is associated with a high reparative capacity (Baker et al. 1955). Bladder muscle tissue is less likely to regenerate in a normal fashion. Both the urothelial and muscle ingrowth are believed to be initiated from the edges of the normal bladder towards the region of the free graft (Baker et al. 1955; Gorham et al. 1989). Usually, however, contracture or resorption of the graft has been evident. The inflammatory response towards the matrix may contribute to the resorption of the free graft.

We hypothesized that building a three-dimensional structure construct in vitro, prior to implantation, would facilitate the eventual terminal differentiation of the cells after implantation in vivo, and would minimize the inflammatory response towards the matrix, thus avoiding graft contracture and shrinkage. In the study above, there was a more aggressive inflammatory reaction in the matrices implanted without cells. Of interest is that the urothelial cell layers appeared normal, even though its underlying matrix was significantly inflamed. We further hypothesized, therefore, (1) that having an adequate urothelial layer from the outset would limit the amount of urine contact with the matrix, and would therefore decrease the inflammatory response, and (2) that the muscle cells were also necessary for bioengineering, being that native muscle cells are less likely to regenerate over the free grafts. Further studies performed in our laboratory confirmed this hypothesis (Oberpenning et al. 1999). Thus, the presence of both urothelial and muscle cells on the matrices we used for bladder replacement appear to be important for successful tissue bioengineering.

In the bladder augmentation study above, a large portion of the native bladders were preserved (Yoo et al. 1998a). When performing studies wherein a large portion of the native bladder is preserved, it is hard to determine if the functional parameters seen (urodynamic findings, such as compliance) are derived from the native bladder itself, or the implanted matrix. In order to better address the functional parameters of tissue-engineered bladders, an animal model was designed which required a subtotal cystectomy (bladder removal) with subsequent replacement with either a cell-seeded or unseeded polymer scaffold (Oberpenning et al. 1999).

A total of 14 beagle dogs underwent a trigone-sparing cystectomy. The animals were randomly assigned to one of three groups. Group A ($n=2$) underwent closure of the trigone without a reconstructive procedure. Group B ($n=6$) underwent reconstruction with an unseeded cell-free bladder-shaped biodegradable polymer. Group C ($n=6$) underwent reconstruction using a bladder-shaped biodegradable polymer that delivered autologous urothelial cells and smooth muscle cells. The cell populations had been separately expanded from a previously harvested autologous bladder biopsy. Preoperative and postoperative urodynamic and radiographic studies were performed serially. Animals were sacrificed at 1, 2, 3, 4, 6, and 11 months postoperatively. Gross, histological, and immunocytochemical analyses were performed (Oberpenning et al. 1999).

The cystectomy-only controls and polymer-only grafts maintained average capacities of 24% and 46% of preoperative values, respectively. An average bladder capacity of 95% of the original pre-cystectomy volume was achieved in the tissue-engineered bladder replacements (Fig. 1). The subtotal cystectomy reservoirs which were not recon-

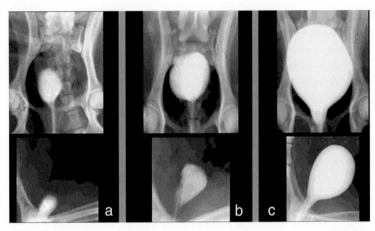

Fig. 1. Radiographic cystograms 11 months after subtotal cystectomy. Subtotal cystectomy bladders retained a small-sized reservoir. Tissue-engineered neobladders showed a normal configuration and a larger capacity than the trigones grafted with polymers only. **a** Subtotal cystectomy only. **b** Polymer-only implants. **c** Tissue-engineered neobladder

structed and polymer-only reconstructed bladders showed a marked decrease in bladder compliance (10% and 42%). The compliance of the tissue engineered bladders showed almost no difference from preoperative values that were measured when the native bladder was present (106%). Histologically, the polymer-only bladders presented a pattern of normal urothelial cells with a thickened fibrotic submucosa and a thin layer of muscle fibers. The retrieved tissue engineered bladders showed a normal cellular organization, consisting of a tri-layer of urothelium, submucosa. and muscle. Immunocytochemical analyses for desmin, alpha actin, cytokeratin 7, pancytokeratins AE1/AE3, and uroplakin III confirmed the muscle and urothelial phenotype. S-100 staining indicated the presence of neural structures. The results from this study showed that it is possible to engineer anatomically and functionally normal bladders using cell-seeded matrices. However, unseeded matrices, without cells, are not adequate for the formation of functionally adequate bladder reservoirs.

4 Urethra

A similar strategy as described above has been used in trying to engineer urethral tissue (Cilento et al. 1996; Chen et al. 1999; Atala et al. 1999). Non-woven meshes of polyglycolic acid have been used experimentally (Cilento et al. 1996). Partial urethrectomies were performed in rabbits and a segment of the polymer mesh with the appropriate diameter was interposed to form the neourethra in each animal. There was no evidence of voiding difficulties or any other complications. Retrograde urethrograms showed no evidence of stricture formation. Histological examination of the neourethras demonstrated complete-re-epithelialization of the polymer implanted sites by day 14 and continued for the entire duration of the study. Polymer fiber degradation was evident 14 days after implantation.

A naturally derived acellular collagen-based tissue substitute developed from donor bladder has also been used for urethral repair. A ventral urethral defect measuring (approximately half of the urethral circumference) was created in ten male rabbits. The acellular collagen matrix was trimmed and used to replace the urethral defect in an onlay fashion (Chen et al. 1999).

Serial urethrograms confirmed the maintenance of a wide urethral caliber without any signs of strictures. Gross examination at retrieval showed normally appearing tissue without any evidence of fibrosis. At retrieval, the distances between the marking sutures placed at the anastomotic margins remained stable, with no distance varying more than 10% in any axis, indicating the maintenance of the initial implant diameter. Histologically, the implanted matrices contained host cell infiltration and generous angiogenesis by 2 weeks after surgery. Minimal infiltration of inflammatory cells was observed initially; however, complete disappearance of these cells was evident by 3 months post-operation. There was no evidence of fibrosis or scarring in the urethras at any of the retrieval time periods.

The presence of a complete transitional cell layer over the graft was confirmed 2 weeks after the repair, and this was consistent throughout the study. The urothelial cell layers stained positively with the broadly reacting anti-pancytokeratins AE1/AE3 in all implants (Fig. 2A). There was no evidence of muscle fibers either at the 2-week or 1-month retrieved implants. Unorganized muscle fiber bundles were evident his-

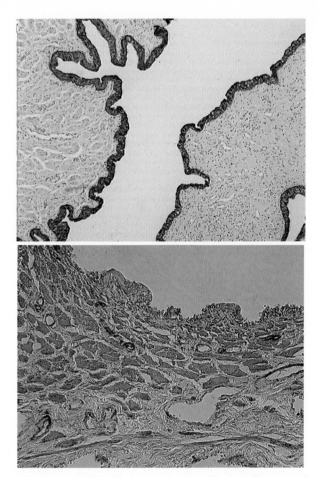

Fig. 2. a Entire urothelial cell layer stained positively with pancytokeratins AE1/AE3 antibodies. Note the native portion of the urethra (*left*) versus the implanted side (*right*). Reduced from ×40. **b** Normally appearing muscle bundles were seen 6 months after implantation. Alpha-actin antibodies, reduced from ×100

tologically 2 months after implantation. The histological patterns suggested that the ingrowth of muscle fibers occurred from all the adjacent native tissue areas, including the ends and sides of the grafts. These findings were confirmed using anti-alpha actin antibodies. Increasing number of organized muscle bundles were observed at 3 months. Normally appearing organized muscle fiber bundles were evident 6 months after implantation (Fig. 2B). These results demonstrated that the acellular collagen matrix could be a useful material for urethral repair in rabbit.

After our experimental experience with the collagen-based acellular matrix, we used the material clinically for urethral reconstruction (Atala et al. 1999). Four patients with a history of prior hypospadias surgery underwent reoperative repair using the collagen-based matrix for urethral reconstruction. The collagen matrix, obtained from donor cadaver bladder, was processed and trimmed to size as needed for each individual patient. The neourethras were created by anastomosing the matrix in an onlay fashion to the urethral plate. The size of the created neourethra ranged from 5 cm to 15 cm. After a 22 -month follow-up, three of the four patients had a successful outcome in regards to their cosmetic appearance and function. One patient who had a 15-cm neourethra created developed a subglanular fistula. These results show that the use of a collagen-based acellular matrix appears to be beneficial for patients with prior hypospadias repair who may lack sufficient genital skin for reconstruction. The acellular collagen-based matrix eliminated the necessity of performing additional surgical procedures for graft harvesting. In addition, operative time, as well as the potential morbidity due to the harvest procedure, was decreased.

From the above studies we have been able to determine that an unseeded matrix can be successfully used in an onlay fashion, and cell-seeded matrices can be successfully used in a tubularized fashion for urethral repair. However, if an unseeded matrix is used in a tubularized fashion, then contracture occurs. These findings suggest that the seeded cells are not essential for successful tissue engineering if the defects are small, such as half the circumference of a urethra, but are necessary if the defects to be repaired are large, such as an entire urethral segment.

5 Kidney

End stage renal failure is a devastating disease which involves multiple organs in affected individuals. Although dialysis can prolong survival for many patients with end-stage renal disease, only renal transplantation can currently restore normal function. Renal transplantation is severely limited by a critical donor shortage. Augmentation of either isolated or total renal function with kidney cell expansion in vitro and subsequent autologous transplantation may be a feasible solution. However, kidney reconstitution using tissue-engineering techniques is a challenging task. The kidney is responsible not only for urine excretion but for several other important metabolic functions in which critical kidney by-products, such as renin, erythropoietin, and Vitamin D, play a large role. We explored the possibility of harvesting and expanding renal cells in vitro and implanting them in vivo in a three-dimensional organization in order to achieve a functional artificial renal unit wherein urine production could be achieved (Atala et al. 1995; Yoo et al. 1996). Studies demonstrated that renal cells can be successfully harvested, expanded in culture, and transplanted in vivo where the single suspended cells form and organize into functional renal structures which are able to excrete high levels of uric acid and creatinine through a yellow urine-like fluid.

Other approaches have also been pursued for renal functional replacement. Polysulphone hollow fibers have been prelined with various extracellular matrix components and seeded with mammalian renal tubular and endothelial cells (Cieslinski et al. 1994). Permselective convective fluid transfer and active transport of salt and water were demonstrated. Using this approach, prototypic biohybrid constructs have been developed which are able to replicate the renal excretory functions. In addition, this system is able to facilitate gene and cell therapies by modifying the cells before seeding.

6 Injectable Therapies

Both urinary incontinence and vesicoureteral reflux are common conditions affecting the genitourinary system, wherein injectable bulking agents can be used for treatment. The goal of several investigators has been to find alternate implant materials which would be safe for human use (Kershen and Atala 1999).

We conducted long-term studies to determine the effect of injectable chondrocytes in vivo as a potential bulking agent (Atala et al. 1993a). We initially determined that alginate, a liquid solution of guluronic and mannuronic acid, embedded with chondrocytes, could serve as a synthetic substrate for the injectable delivery and maintenance of cartilage architecture in vivo. Alginate undergoes hydrolytic biodegradation and its degradation time can be varied depending on the concentration of each of the polysaccharides. The use of autologous cartilage for the treatment of vesicoureteral reflux in humans would satisfy all the requirements for an ideal injectable substance. A biopsy of the ear could be easily and quickly performed, followed by chondrocyte processing and endoscopic injection of the autologous chondrocyte suspension for the treatment reflux.

Chondrocytes can be readily grown and expanded in culture. Neocartilage formation can be achieved in vitro and in vivo using chondrocytes cultured on synthetic biodegradable polymers (Atala et al. 1993a). In our experiments, the cartilage matrix replaced the alginate as the polysaccharide polymer underwent biodegradation. We then adapted the system for the treatment of vesicoureteral reflux in a porcine model (Atala et al. 1994).

Six mini-swine underwent bilateral creation of reflux. All six were found to have bilateral reflux without evidence of obstruction at 3 months following the procedure. Chondrocytes were harvested from the left auricular surface of each mini-swine and expanded with a final concentration of 50×10^6 viable cells per animal. The animals underwent endoscopic repair of reflux with the injectable autologous chondrocyte solution on the right side only.

Serial cystograms showed no evidence of reflux on the treated side and persistent reflux in the uncorrected control ureter in all animals. All animals had a successful cure of reflux in the repaired ureter without evidence of hydronephrosis on excretory urography. The harvested ears had evidence of cartilage regrowth within 1 month of chondrocyte retrieval. At the time of sacrifice, gross examination of the bladder injection site showed a well-defined rubbery to hard cartilage structure in the subureteral region. Histological examination of these specimens showed evidence of normal cartilage formation. The polymer gels were progressively replaced by cartilage with increasing time. Aldehyde fuschin-alcian blue staining suggested the presence of chondroitin sulfate. Microscopic analyses of the tissues surrounding the injection site showed no inflammation. Tissue sections from the bladder, ureters, lymph nodes, kidneys, lungs, liver, and spleen showed no evidence of chondrocyte or alginate migration, or granuloma formation. These studies showed that chondrocytes can be easily harvested and combined with alginate in vitro, the suspension can be easily injected cystoscopically, and the elastic cartilage tissue formed is able to correct vesicoureteral reflux without any evidence of obstruction (Atala et al. 1994).

Using the same line of reasoning as with the chondrocyte technology, our group investigated the possibility of using autologous muscle cells (Cilento and Atala 1995). In vivo experiments were conducted in minipigs and reflux was successfully corrected. In addition to its use for the endoscopic treatment of reflux and urinary incontinence, the system of injectable autologous cells may also be applicable for the treatment of other medical conditions, such as rectal incontinence, dysphonia, plastic reconstruction, and wherever an injectable permanent biocompatible material is needed.

Recently, the first human application of cell-based tissue-engineering technology for urological applications has occurred with the injection of chondrocytes for the correction of vesicoureteral reflux in children and for urinary incontinence in adults. The clinical trials are currently ongoing.

7 Testis

Leydig cells are the major source of testosterone production in males. Patients with testicular dysfunction require androgen replacement for somatic development. Conventional treatment for testicular dysfunction consists of periodical IM injections of chemically modified testosterone or, more recently, of skin-patch applications. However, long-term nonpulsatile testosterone therapy is not optimal and can cause multiple problems, including erythropoiesis and bone density changes.

A system was designed wherein Leydig cells were microencapsulated for controlled testosterone replacement. Microencapsulated Leydig cells offer several advantages, such as serving as a semipermeable barrier between the transplanted cells and the host's immune system, as well as allowing for the long-term physiological release of testosterone.

Purified Leydig cells were isolated, characterized, suspended in an alginate solution, and extruded through an air jet nozzle into a 1.5% $CaCl_2$ solution were they gelled; and were further coated with 0.1% poly-L-lysine. The encapsulated cells were pulsed with human chorionic gonadotropin (HCG) every 24 h. The medium was sampled at different time points after HCG stimulation and analyzed for testosterone production. Cell viability was confirmed daily. The encapsulated Leydig cells were injected in castrated animals and serum testosterone was measured serially. The castrated animals receiving the microencapsulated cells were able to maintain testosterone levels long-term (Machluf et al. 1998). These studies suggest that microencapsulated Leydig cells may be able to replace or supplement testosterone in situations were anorchia or testicular failure is present. A similar system is currently being applied for estrogen.

8 Genitalia

A large number of congenital and acquired abnormalities of the genitourinary system, including ambiguous external genitalia, the exstrophy–epispadias complex and impotence, would benefit from the availability of transplantable, autologous corpus cavernosum tissue for use in reconstructive procedures. Given the major structural and functional importance of corpora cavernosal tissue, it is clear that the availability of autologous corporal smooth muscle tissue for use in reconstructive procedures would be of great clinical utility, facilitating enhanced cosmetic results, while providing the possibility of de novo, functional erectile tissue.

Experiments performed in our laboratory were designed to determine the feasibility of using cultured human corporal smooth muscle cells seeded onto biodegradable matrix scaffolds for the formation of corpus

cavernosum muscle in vivo. Primary cultures of human corpus cavernosum smooth muscle cells were derived from operative biopsies obtained
during penile prosthesis implantation and vaginal resection. Cells were
maintained in continuous multilayered cultures, seeded onto polymers
of nonwoven polyglycolic acid, and implanted subcutaneously in
athymic mice. Animals were sacrificed at various time points after
surgery and the implants were examined via histology, immunocyto-
chemistry, and Western blot analysis (Kershen et al. 1998).

Corporal smooth muscle tissue was identified grossly, and histologically at the time of sacrifice. Intact smooth muscle cell multilayers were
observed growing along the surface of the polymers throughout all
retrieved time points. There was evidence of early vascular ingrowth at
the periphery of the implants by 7 days. By 24 days, there was evidence
of polymer degradation. Smooth muscle phenotype was confirmed im-
munocytochemically and by Western blot analysis with antibodies to
alpha-smooth muscle actin.

Further studies were performed wherein corpora cavernosal muscle
cells were co-cultured with endothelial cells. The co-cultured cells were
seeded on polymers and implanted in vivo. At retrieval, by 6 weeks,
there was tissue organization similar to normal corpora (Park et al.
1999). These studies provided the first evidence that cultured human
corporal smooth muscle cells could be used in conjunction with biodegradable polymer scaffolds to create corpus cavernosum tissue de novo.

Currently, the principal method of reconstructing a phallus when
insufficient tissue is present, is to utilize silicone rigid prostheses. Although silicone penile prostheses have been an accepted treatment mo-
dality since the 1970s, issues with biocompatibility remain (Nukui et al.
1997). Creation of natural penile prostheses composed of vascularized
autologous tissue may be advantageous. We investigated the possibility
of creating a natural phallic prosthesis consisting of autologous chondrocytes, which if biocompatible and elastic, could be used in patients
who require genital reconstruction.

Cartilage was harvested from the articular surface of calf shoulders.
Chondrocytes were isolated, grown, and expanded in vitro. The cells
were seeded onto preformed cylindrical polyglycolic acid polymer rods
at a concentration of 50×10^6 chondrocytes/cm^3. Cell-polymer scaffolds
were implanted in vivo. Each mouse had two implantation sites consisting of a polymer scaffold seeded with chondrocytes and a control

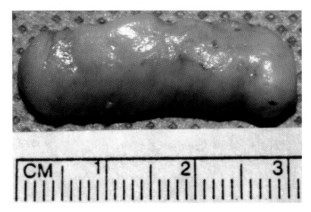

Fig. 3. Retrieved implant 6 months after implantation contains well-formed, milky white rod-shaped cartilaginous structure

(polymer alone). The engineered rods were retrieved at 1, 2, 4, and 6 months after implantation. Stress-relaxation studies to measure biomechanical properties, including compression, tension, and bending, were performed on the retrieved structures. Histological analyses were performed with hematoxylin and eosin, aldehyde fuschin-alcian blue, and toluidine blue staining (Yoo et al. 1998).

Gross examination showed the presence of well-formed milky white rod-shaped solid cartilage structures which were approximately the same size as the initial implant (Fig. 3). A series of stress relaxation tests were performed in order to determine whether the engineered cartilage rods possessed the mechanical properties required to maintain penile rigidity. Biomechanical analyses of all specimens demonstrated similar patterns. The compression studies showed that the retrieved cartilage rods were able to withstand high degrees of pressure. A ramp compression speed of 200 µm/s, applied to each cartilage rod up to 2,000 µm in distance, resulted in 3.8 kg of resistance. The tension relaxation studies demonstrated that the retrieved cartilage rods were able to withstand stress and were able to return to their initial state while maintaining their biomechanical properties. A ramp tension speed of 200 µm/s applied to each cartilage rod created a tensile strength of 2.2 kg, which physically lengthened the rods an average of 0.48 cm. Relaxation of tension at the same speed resulted in retraction of the cartilage rods to their initial state. The 5 cycles of bending studies performed at two different speeds showed that the engineered cartilage rods were durable, malleable, and were able to retain their mechanical properties. None of the rods were ruptured during the biomechanical stress relaxation studies, which showed that the cartilage structures were readily elastic and could withstand high degrees of pressure. Histochemical analyses with hematoxylin and eosin, aldehyde fuschin-alcian blue, and toluidine blue staining demonstrated the presence of mature and well-formed chondrocytes in all the implants. There was no evidence of cartilage formation in the controls.

In a subsequent study, autologous cartilage seeded rods were implanted into rabbit corporas. The scaffolds were able to form cartilage rods in vivo in the corpora. The engineered penile prostheses were stable, without any evidence of infection or erosion (Yoo et al. 1999). These preliminary studies indicate that creation of a penile prosthesis composed of chondrocytes may be achieved using biodegradable polymer scaffolds as a cell delivery vehicle. The engineered tissue forms a cartilaginous structure which resists high pressures. The use of an autologous system would preclude an immunological reaction. This technology could be useful in the future for the creation of a bio-compatible malleable prosthesis for patients undergoing penile reconstruction.

9 Fetal Applications

The prenatal diagnosis of fetal abnormalities is now more prevalent than ever before. Prenatal ultrasonography allows for a thorough survey of fetal anatomy. For example, the absence of bladder filling, a mass of echogenic tissue on the lower abdominal wall or a low set umbilicus during prenatal sonographic examination may suggest the diagnosis of bladder exstrophy. These findings and the presence of intraluminal intestinal calcifications suggest the presence of a cloacal malformation.

The natural consequence of the evolution in prenatal diagnosis led to the use of intervention before birth to reverse potentially life-threatening processes. However, the concept of prenatal intervention itself is not limited to this narrow group of indications. A prenatal, rather than a postnatal diagnosis of urological conditions, such as exstrophy, may be beneficial under certain circumstances. There is now a renewed interest in performing a single-stage reconstruction in some patients with bladder exstrophy. Limiting factors for following a single or multiple-stage approach may include the findings of a small, fibrotic bladder patch without either elasticity or contractility, or a hypoplastic bladder.

There are several strategies which may be pursued, using today's technological and scientific advances, which may facilitate the future prenatal management of patients with bladder disease. Having a ready supply of urologically associated tissue for surgical reconstruction at birth may be advantageous. Theoretically, once the diagnosis of bladder exstrophy is confirmed prenatally, a small bladder and skin biopsy could be obtained via ultrasound guidance. These biopsy materials could then be processed and the different cell types expanded in vitro. Using tissue engineering techniques developed at our center and described previously, reconstituted bladder and skin structures in vitro could then be readily available at the time of birth for a one-stage reconstruction, allowing for an adequate anatomical and functional closure.

Toward this end, we conducted a series of experiments using fetal lambs (Fauza et al. 1998). Bladder exstrophy was created surgically in ten 90–95-day gestation fetal lambs. The lambs were randomly divided into two groups of five. In group I, a small fetal bladder specimen was harvested via fetoscopy. The bladder specimen was separated and muscle and urothelial cells were harvested and expanded separately under sterile conditions in a humidified 5% CO_2 chamber, as previously described. Seven to ten days prior to delivery, the expanded bladder muscle cells were seeded on one side and the urothelial cells on the opposite side of a 20-cm^2 biodegradable polyglycolic acid polymer scaffold. After delivery, all lambs in group I had surgical closure of their bladder using the tissue-engineered bladder tissue. No fetal bladder harvest was performed in the group II lambs, and bladder exstrophy closure was performed using only the native bladder. Cystograms were performed 3 and 8 weeks after surgery. The engineered bladders were more compliant ($p=.01$) and had a higher capacity ($p=.02$) than the native bladder closure group. Histological analysis of the engineered tissue showed a normal histological pattern, indistinguishable from native bladder at 2 months (Fauza et al. 1998). Similar prenatal studies were performed in lambs, engineering skin for reconstruction at birth (Fauza et al. 1998). Other fetal tissues, such as cartilage, corpora cavernosa, and skeletal muscle can also be harvested and expanded in the same manner. Similar studies addressing these tissues are now in progress in our laboratory.

10 Gene Therapy

Based on the feasibility of tissue engineering techniques in which cells seeded on biodegradable polymer scaffolds form tissue when implanted in vivo, the possibility was explored of developing a neo-organ system for in vivo gene therapy (Yoo and Atala 1997).

In a series of studies conducted in our laboratory, human urothelial cells were harvested, expanded in vitro and seeded on biodegradable polymer scaffolds. The cell–polymer complex was then transfected with PGL3-*luc*, pCMV-*luc* and pCMVβ-*gal* promoter-reporter gene constructs. The transfected cell-polymer scaffolds were then implanted in vivo and the engineered tissues were retrieved at different time points after implantation. Results indicated that successful gene transfer could be achieved using biodegradable polymer scaffolds as a urothelial cell delivery vehicle. The transfected cell/polymer scaffold formed organ-like structures with functional expression of the transfected genes (Yoo and Atala 1997).

This technology is applicable throughout the spectrum of diseases which may be manageable with tissue engineering. For example, one can envision the use effective in vivo gene delivery through the ex vivo transfection of tissue-engineered cell/polymer scaffolds for the genetic modification of diseased corporal smooth muscle cells harvested from impotent patients. Studies of human corpus cavernosum smooth muscle cells have suggested that cellular overproduction of the cytokine transforming growth factor (TGF)-1 may lead to the synthesis and accumulation of excess collagen in patients with arterial insufficiency resulting in corporal fibrosis. Prostaglandin E1 (PGE1) was shown to suppress this effect in vitro. Theoretically, the in vitro genetic modification of corporal smooth muscle cells harvested from an impotent patient, resulting in either a reduction in the expression of the TGF-1 gene, or the overexpression of genes responsible for PGE1 production, could lead to the resumption of erectile functionality once these cells were used to repopulate the diseased corporal bodies.

11 Conclusion

Tissue engineering efforts are currently being undertaken for every type of tissue and organ within the urinary system. Most of the effort expended to engineer genitourinary tissues has occurred within the last decade. Tissue engineering techniques require expertise in growth factor biology, a cell-culture facility designed for human application, and personnel who have mastered the techniques of cell harvest, culture, and expansion. Polymer scaffold design and manufacturing resources are essential for the successful application of this technology. In order to apply these engineering techniques to humans, further studies need to be performed in many of the tissues described.

The first human application of cell-based tissue engineering technology for urological applications has occurred at our institution with the injection of autologous cells for the correction of vesicoureteral reflux in children. These clinical trials are currently ongoing. The same technology has been recently expanded to treat adult patients with urinary incontinence. Furthermore, trials involving urethral tissue replacement using processed collagen matrices are in progress and bladder replacement using tissue engineering techniques are currently being arranged. Recent progress suggests that engineered urological tissues may have clinical applicability in the future.

References

Amiel GE, Atala A (1999) Current and future modalities for functional renal replacement. Urol Clin 26:235–246

Atala A (1995) Commentary on the replacement of urologic associated mucosa. J Urol 156:338–339

Atala A (1997) Tissue engineering in the genitourinary system. In: Atala A, Mooney D (eds) Tissue engineering. Birkhauser, Boston, Chap 8

Atala A (1998) Autologous cell transplantation for urologic reconstruction. J Urol 159:2–3

Atala A (1999) Future perspectives in reconstructive surgery using tissue engineering. Urol Clin 26:157–165

Atala A, Vacanti JP, Peters CA, Mandell J, Retik AB, Freeman MR (1992) Formation of urothelial structures in vivo from dissociated cells attached to biodegradable polymer scaffolds in vitro. J Urol 148:658–662

Atala A, Cima LG, Kim WS, Page KT, Vacanti JP, Retik AB, Vacanti CA (1993a) Injectable polymers seeded with chondrocytes as a therapeutic approach. J Urol 150:745–747

Atala A, Freeman MR, Vacanti JP, Shepard J, Retik AB (1993b) Implantation in vivo and retrieval of artificial structures consisting of rabbit and human urothelium and human bladder muscle. J Urol 150:608–612

Atala A, Cima LG, Kim W, Paige KT, Vacanti JP, Retik AB, Vacanti CA (1993c) Injectable alginate seeded with chondrocytes as a potential treatment for vesicoureteral reflux. J Urol 150:745–747

Atala A, Kim W, Paige KT, Vacanti CA, Retik AB (1994) Endoscopic treatment of vesicoureteral reflux with chondrocyte-alginate suspension. J Urol 152:641–644

Atala A, Schlussel RN, Retik AB (1995) Renal cell growth in vivo after attachment to biodegradable polymer scaffolds. J Urol [Suppl] 153:4

Atala A, Guzman L, Retik A (1999) A novel inert collagen matrix for hypospadias repair. J Urol 162:1148–1151

Baker R, Kelly T, Tehan T, Putman C, Beaugard E (1955) Subtotal cystectomy and total bladder regeneration in treatment of bladder cancer. J Am Med Ass 168:1178

Chen F, Yoo JJ, Atala A (1999) Acellular collagen matrix as a possible "off the shelf" biomaterial for urethral repair. Urol 54:407–410

Cieslinski, DA, Humes HD (1994) Tissue engineering of a bioartificial kidney. Biotech Bioeng 43:678

Cilento BG, Atala A (1995) Treatment of reflux and incontinence with autologous chondrocytes and bladder muscle cells. Dialogues Pediatr Urol 18(11):2–3

Cilento BG, Freeman MR, Schneck FX, Retik AB, Atala A (1994) Phenotypic and cytogenetic characterization of human bladder urothelia expanded in vitro. J Urol 152:655–670

Cilento BG, Retik AB, Atala A (1996) Urethral reconstruction using a polymer scaffolds seeded with urothelial and smooth muscle cells. J Urol [Suppl] 155:5

Fauza DO, Fishman S, Mehegan K, Atala A (1998) Videofetoscopically assisted fetal tissue engineering: bladder augmentation. J Pediatr Surg 33:7–377

Folkman J, Hochberg MM (1973) Self-regulation of growth in three dimensions. J Exp Med 138:745–753

Gorham SD, French DA, Shivas AA, Scott R (1989) Some observations on the regeneration of smooth muscle in the repaired urinary bladder of the rabbit. Eur Urol 16:440–443

Kershen RT, Atala A (1999) New advances in injectable therapies for the treatment of incontinence and vesicoureteral reflux. Urol Clin 26:81–94

Kershen RT, Yoo JJ, Moreland RB, Krane RJ, Atala A (1998) Novel system for the formation of human corpus cavernosum smooth muscle tissue in vivo. J Urol [Suppl] 159:156

Machluf M, Atala A (1998) Emerging concepts for tissue and organ transplantation. Graft 1:31–37

Machluf M, Boorjian S, Caffaratti J, Kershen R, Atala A (1998) Microencapsulation of Leydig cells: a new system for the therapeutic delivery of testosterone. Pediatrics 102S:32

Nukui F, Okamoto S, Nagata M, Kurokawa J, Fukui J (1997) Complications and reimplantation of penile implants. Int J Urol 4:52–54

Oberpenning FO, Meng J, Yoo JJ, Atala A (1999) De novo reconstitution of a functional urinary bladder by tissue engineering. Nature Biotech 17:149–155

Park HJ, Kershen R, Yoo JJ, Atala A (1999) Reconstitution of human corporal smooth muscle and endothelial cells in vivo. J Urol 162:1106–1109

Yoo JJ, Atala A (1997) A novel gene delivery system using urothelial tissue engineered neo-organs. J Urol 158:1066–1070

Yoo JJ, Satar N, Retik AB, Atala A (1995) Ureteral replacement using biodegradable polymer scaffolds seeded with urothelial and smooth muscle cells. J Urol [Suppl] 153:4

Yoo JJ, Ashkar S, Atala A (1996) Creation of functional kidney structures with excretion of urine-like fluid in vivo. Pediatrics 98S:605

Yoo JJ, Meng J, Oberpenning F, Atala A (1998a) Bladder augmentation using allogenic bladder submucosa seeded with cells. Urol 51:221

Yoo JJ, Lee I, Atala A (1998b) Cartilage rods as a potential material for penile reconstruction. J Urol 160:1164–1168

Yoo JJ, Park HJ, Lee I, Atala A (1999) Autologous engineered cartilage rods for penile reconstruction. J Urol 162:1119–1121

Index

Ernst Schering Research Foundation Workshop

Editors: Günter Stock
Monika Lessl

This series will be available on request from
Ernst Schering Research Foundation, 13342 Berlin, Germany